The New Worlds
Extrasolar Planets

Fabienne Casoli and Thérèse Encrenaz

The New Worlds

Extrasolar Planets

Springer

Published in association with
Praxis Publishing
Chichester, UK

Dr Fabienne Casoli
Institut d'Astrophysique Spatiale
    (Orsay) CNRS and Université
    Paris-Sud 11
Orsay
France

Dr Thérèse Encrenaz
Laboratoire d'études spatiales et
    d'instrumentation en astrophysique
    (LESIA)
Observatoire de Paris
Meudon
France

Original French edition: *Planètes extrasolaires: Les nouveaux mondes*
Published by © Éditions Belin 2005
Ouvrage publié avec le concours du Ministère français chargé de la culture – Centre
National du Livre
This work has been published with the help of the French Ministère de la Culture –
Centre National du Livre

Translator: Bob Mizon, 38 The Vineries, Colehill, Wimborne, Dorset, UK

SPRINGER–PRAXIS BOOKS IN POPULAR ASTRONOMY
SUBJECT *ADVISORY EDITOR*: John Mason B.Sc., M.Sc., Ph.D.

ISBN 10: 0-387-44906-X Springer Berlin Heidelberg New York

Springer is a part of Springer Science + Business Media (*springer.com*)

Library of Congress Control Number: 2006932709

Cover design: Jim Wilkie
Copy editing: R. A. Marriott
Typesetting: BookEns Ltd, Royston, Herts., UK

Printed in Germany on acid-free paper

Contents

Authors' preface

Authors' preface

Exoplanet, extrasolar planet, exoEarth, exoJupiter: neologisms still absent from many dictionaries. These terms are, however, current among astronomers, and are heard in their answers to a question already two millennia old: are there planets like ours elsewhere in the Universe?

Greek atomists such as Epicurus were convinced of the existence of an infinite number of solar systems like our own, but it was only in 1995 that a real answer began to emerge. An extrasolar planet had been detected... a planet orbiting another star... a star like the Sun. So, the solar system was not unique! By mid-2006 more than 200 giant exoplanets had been discovered. At this rate of discovery it seems that Earth-like planets may be found within a decade.

The discovery of exoplanets held some surprises, in that they exhibited very different characteristics from what might have been expected. Although most of them are gas giants of masses comparable to Jupiter's mass, as a result of the rather insensitive nature of current detection methods, why are they from ten to fifty times closer to their stars than is Jupiter? How were these 'hot Jupiters' formed?

Another surprise about exoplanets is that many of them have very elliptical orbits, while the planets of the solar system have much more circular orbits. Could the solar system therefore be an exceptional case? Will we have to revise our ideas about its formation?

The most exciting aspect of these discoveries of exoplanets is the search for exoEarths – planets similar in mass to ours, orbiting not far from the parent star: indeed, planets upon which life might well develop. The challenge for the next twenty years will be to identify these exoEarths and seek signs there of some form of life. Worldwide, more than forty research programmes now exist in this field, and as many more are planned. Searching for other worlds is now a beacon activity among the scheduled tasks for European and American large Earth-based telescopes and space missions.

The methods used to detect exoplanets and to analyse the results are very varied, and push modern instruments to their limits. The study of exoEarths will require extraordinary telescopes, using innovative techniques, most of which have only recently been introduced. As far as space missions are concerned, the COROT satellite for searching extrasolar planets was launched in December 2006, there are several missions which are dedicated to the search for water on

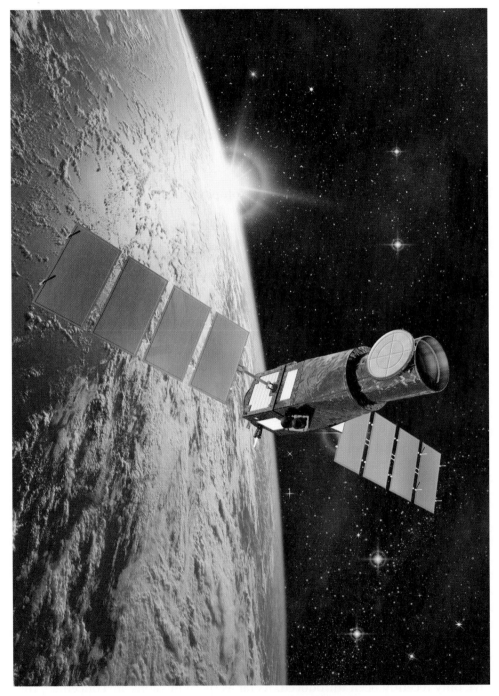

The French COROT mission, launched by the CNES (National Centre for Space Studies) in 2006.

Mars, and later missions will be searching for traces of any life which may once have emerged on that planet.

The detection and study of these extrasolar worlds has become a very dynamic and fast-moving aspect of astronomy – a situation which presents a certain challenge to the authors of any up-to-the-minute book on the subject! There are, at the moment, no hard and fast conclusions to be drawn. In the summer of 2004 it seemed reasonable to state that direct imaging of an extrasolar planet would not be possible for at least a decade – yet the first probable image of such a planet was published in September 2004 and confirmed in April 2005! The number of exoplanets detected increases all the time, and the record for 'the least massive planet yet discovered' is regularly broken. In 2006 the first cold exoEarth, with a mass almost six times that of Earth, was detected in orbit around a small M-type star.

Nevertheless, we decided to accept the challenge of exploring this subject. The major results obtainable with current instrumentation are understood, and the main questions are tabled. The next step forward will be taken by spaceborne instruments onboard the Franco-European COROT mission and the American Kepler mission (2008). What surprises might extrasolar planets yet hold in store for us?

1
Extrasolar planets: the Holy Grail of astronomers

As early as the third century BC the Greek philosopher Epicurus (341–270 BC) affirmed the existence of an infinite number of worlds. In a letter to Herodotus he wrote: 'Epicurus to Herodotus, greetings... The Universe is infinite. Moreover, there is an infinite number of worlds, some like this world, others unlike it. Also, atoms are infinite in number, as has just been proved, and they are carried ever further in their courses. Now, the atoms from which a world might be made, or from which a world is already made, have not all been expended on only one world, or on a finite number of worlds, whether they be like or unlike each other. Hence there is nothing to oppose the idea of an infinity of worlds.'

The Orion Nebula – a site of star formation – imaged by the Hubble Space Telescope in 1991.

1.1 TWELVE YEARS OF DISCOVERIES AND SURPRISES

Astronomers find planets in unexpected places

Are we alone in the Universe? This is a question which divided philosophers as long ago as the age of the ancient Greeks. Epicurus thought that the number of worlds was infinite, while Aristotle believed that the Earth was unique and at the centre of the Universe. Twenty-four centuries of debate and discussion went by before astronomers were able to prove Aristotle wrong! The Earth is not unique, and since 1995 we have known that there also exist planets orbiting stars like the Sun: the extrasolar planets (exoplanets). More than 200 of these worlds are now known, and the number increases monthly. None of these planets really resembles the Earth, but astronomers are sure that exoEarths will be found as soon as appropriate instruments become available.

So, there are other solar systems. Their exploration by astronauts cannot be accomplished for centuries to come, but by 2010, due to space missions still at the planning stage, we should know if there are rocky, Earth-like planets out there. They may orbit their parent stars at distances not too near, and not too far, for life to have gained a foothold.

Twenty years from now we could have the necessary instrumentation to be able to search these exoEarths for signs of life: an atmosphere containing oxygen, or the signature of chlorophyll.

So many discoveries – and surprises – have come in quick succession since 1992, when the only solar system we knew of was our own! This was the year in

Aristotle (348–322 BC) believed the Earth, at the centre of the Universe, to be unique.

Epicurus (341–270 BC) envisaged the existence of multiple worlds.

The Earth – a life-bearing planet with oceans, land masses, clouds and polar caps (seen here from Apollo 12) – bears all the hallmarks of a planet where life has developed. We are a long way from being able to say the same of exoplanets.

Another Earth, 10 light-years away? A synthetic image of an exoEarth 10 light-years away, as it might be seen by an armada of 150 3-metre space telescopes travelling 150 km apart – a project still in the files of the space agencies!

which radio astronomers sprang the first surprise: Earth-sized planets orbiting a pulsar – a star at the end of its life cycle. This was absolutely the last place where one might have expected planets to be, since surely no planet could have survived the supernova explosion which results in a pulsar.

In 1995 there was another surprise when a Swiss team announced the discovery of an exoplanet comparable in mass to Jupiter. This planet orbits a fairly ordinary Sun-like star called 51 Pegasi (in the constellation of Pegasus – the number referring to John Flamsteed's eighteenth-century catalogue of naked-eye stars in order of right ascension, from west to east).

A third surprise awaited astronomers: 51 Pegasi b (as this planet is sometimes designated) orbits its star in little more than 4 days, compared with the 12 years it takes Jupiter to complete one orbit of the Sun. This means that 51 Pegasi b lies at a distance twenty times less than that of the Earth from the Sun, or an eighth of the distance of Mercury from the Sun. It is easy to imagine that it must be quite a hot world!

Since 1995 the rate at which discoveries have been made has been remarkable, and the number of known exoplanets increases month by month. By July 2006 more than 200 such planets, orbiting more than 170 stars, were listed. With the exception of the three planets associated with the pulsar PSR 1257+12, they are nearly all heavyweights, with masses between 0.02 and 14 times that of Jupiter. Thus we talk of exoJupiters, exoSaturns and exoNeptunes; but exoEarths still elude us, as current techniques cannot detect them near ordinary stars.

The quest for the first Earth-like planet capable of harbouring some kind of life is driving some very ambitious and difficult programmes of research. Early observations depended on instruments such as the 2-metre telescope at the Haute-Provence Observatory, used to discover 51 Pegasi b; but in future programmes, much larger instruments will have to be utilised. There is talk of Earth-based telescopes with apertures of 30–50 metres; and in space, flotillas comprising perhaps dozens of telescopes. There are many ideas for programmes of which the feasibility remains to be proved... but the difficulty of the task reflects the value of the prize.

1.2 FIRST CLUES

1992 and 1995: the first extrasolar planets

On 5 October 1995, two Swiss astronomers – Michel Mayor and Didier Queloz – announced their discovery of an extrasolar planet in orbit around a Sun-like star, 51 Pegasi.

Another false alarm? Since the mid-twentieth century, reports of such sightings had, like stories of sea serpents, been bandied about, only to be discounted by other research teams. Results could not be reproduced, and each 'new planet' vanished in a puff of smoke...

Undaunted in the face of controversy, a few teams of researchers ploughed on doggedly. They were confident that their equipment was up to the task of detecting the presence of large planets of the size of Jupiter. Of course, there was no question of actually 'seeing' such a body: imaging an exceedingly faint planet orbiting a star billions of times brighter than itself is truly impossible. One might just as well try to detect a candle placed alongside a lighthouse, and, for fair comparison, the lighthouse would have to be observed from hundreds of kilometres away!

Detection programmes have therefore to rely on methods involving the minute perturbations induced in a star's motion by the presence of a nearby planet, involving the displacement of a star's apparent position compared with the positions of other stars, or modifications in a star's velocity within our galaxy. These perturbations are regular, with a period reflecting that of the revolution of the planet around the star.

In the case of Jupiter, its presence causes the velocity of the Sun through our galaxy to vary by 13 m/s over a 12-year period – a variation detectable with the best spectrometers since the 1990s.

Moreover, far from having detected planets around ordinary stars, astronomers have detected two around a late-stage star: a pulsar. A pulsar is a star which emits radio 'bursts' at very regular intervals. They are neutron stars – the remnant of the explosion of a star which is much more massive than the Sun: a supernova. In 1992 astronomers detected two planets a little larger than the Earth orbiting

The Crab Nebula – the remnant of a supernova explosion observed in 1054 and reported by Chinese astronomers. At its centre lies a pulsar – a very dense neutron star in rapid rotation (30 times per second). The nebula is all that remains of the rest of the progenitor star. Matter ejected during the explosion forms reddish filaments, and the cloud is expanding at around 1,500 km/s. The first two exoplanets to be found were detected near a pulsar created by a supernova explosion. These planets certainly formed after the explosion; but it is difficult to imagine how a planetary system could survive the enormous burst of energy of such an event. It is not known whether the Crab pulsar possesses any planetary companions. (Photograph courtesy CFHT, J.-C. Cuillandre.)

the pulsar PSR 1257+12. The explosion of a supernova is an event of unimaginable violence, releasing vast tides of energy into space. A shock wave would pass through its planetary system at velocities of several thousand kilometres per second. It is impossible to conceive how planets could survive a hurricane of this nature. However, the data were irrefutable: the pulsar PSR 1257+12 really did possess two planets, or perhaps three (the presence of the third planet being confirmed later). This most unexpected discovery led optimistic researchers to conclude that if planets can be found in such an inhospitable environment, how much more probable would it be that they could be found near ordinary stars?

So, the situation in 1995 was that, apart from the nine planets orbiting the Sun, two planets had been found near a pulsar, and a few teams were searching intently for other planetary bodies. At that time, few people would have bet on the likelihood of such a plethora of discoveries during the following years... which demonstrates that the course of scientific discovery is not always easy to predict.

1.3 THE SOLAR SYSTEM: AN ATYPICAL PLANETARY SYSTEM?

The astronomers sought – but did not find

In order to find exoplanets it is useful to know at least something of their characteristics. Because direct imaging of them is impossible with current instruments, the method employed relies on precise observations of the motion of a star and the detection of the regular perturbations induced in its motion caused by an orbiting planet. The nature of these perturbations depends on the mass of the planet compared with that of the star, on the planet's orbital distance, and on the shape of its orbit. These criteria are estimated in order to determine observational strategies. For example: if we are searching for a planet at the same distance from the star as Jupiter is from the Sun, we know that the period of revolution will be about 12 years. Daily observations will therefore be of little use, but a schedule will need to last for several years.

Naturally enough, all the researchers took the solar system as their model: larger planets on the outside, and smaller planets nearer to the star. The Sun's family of planets can be divided into two classes. The first group comprises rocky planets with an obvious surface, such as Mercury, Venus, the Earth and Mars: the 'terrestrial' planets. The second group comprises the four giants, Jupiter, Saturn, Uranus and Neptune, which are much more massive than the terrestrials, and consist mostly of hydrogen and helium. Until August 2006 Pluto was the ninth planet of the solar system, but with the decision of the General Assembly of the International Astronomical Union it is now classed as a 'dwarf planet'. It is one of the 'trans-Neptunian objects' – small bodies of rock and ice moving far from the Sun beyond the orbit of Neptune.

A terrestrial planet: Mars. The fourth planet from the Sun is half the size of the Earth. It has a rocky surface and a thin atmosphere, 95% of which is carbon dioxide. Mars is accompanied by two tiny satellites: Phobos and Deimos.

A gas giant: Jupiter – the king of the eight planets of the solar system. The diameter of Jupiter is only ten times less than that of the Sun, and its thick hydrogen–helium atmosphere surrounds an icy core about ten times the mass of the Earth. Jupiter now has more than sixty recorded satellites – the most famous being the four Galileans – Io, Europa, Ganymede and Callisto – named after the Italian astronomer Galileo Galilei, who was the first to observe them with a telescope.

The two classes of planets differ not only in their natures, but also in their distances from the Sun. Terrestrials circle near the Sun, and giants further away. The more distant they are, the longer it takes them to revolve around it. Of the inner, terrestrial planets, the orbit of Mercury, nearest the Sun, lasts 88 days, while that of Mars lasts 687 days. The orbital periods of the giant, outer planets range from 12 years (Jupiter) to 165 years (Neptune). All these orbits are more or less circular, except that of Mercury, which is a little more elliptical.

Armed with this knowledge to guide them into the unknown, the planet hunters of the 1990s thought they knew what they were looking for: exoJupiters and exoSaturns with circular orbits and periods of more than 10 years. Their instruments were incapable of detecting less massive planets. The observational strategies they evolved reflected these expectations. However, although the first exoplanet to be detected near a Sun-like star had a mass 0.47 times that of Jupiter, it orbited its star in only 4.2 days! This was such an unexpected discovery that one of the two teams then investigating exoplanets, led by American astronomer Geoff Marcy, missed the object. They thought they would have to wait for several years for a positive detection, and had not begun to examine their data just weeks into the project.

Orbits in the solar system

The eight planets move in ellipses around the Sun, which lies not at the centre C, but at one of the foci F. Most of the planets of the solar system have only slightly elliptical orbits, and are near-circular (a circle being a particular case where C and F are the same).

The two characteristics of an ellipse are its eccentricity – with values between 1 and 0, where 0 denotes a circle and values approaching 1 denote a very flattened ellipse – and the semimajor axis (a), which is the longest distance between the centre and a point on the ellipse.

The further a planet from the Sun, the longer it takes to perform a complete revolution. The period of revolution T of a planet depends on the semimajor axis a, according to Kepler's third law. If T is expressed in years and a in Astronomical Units (AU, the mean Sun–Earth distance – 149,600,000 km), then $a^3/T^2 = 1$. In any planetary system, a^3/T^2 is proportional to the mass of the central star.

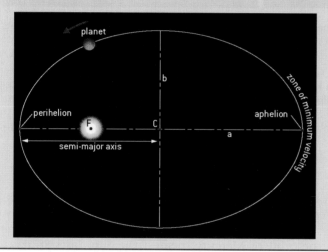

Definition of the semimajor axis of an ellipse with centre C and focus F.

1.4 THE BIG QUESTION

Discoveries bringing more questions than answers

Ten years on, researchers have still not quite come to terms with their surprise of 1995. Although discoveries of exoplanets provide fresh perspectives on the search for life in the Universe, it cannot be denied that these planets are very different from those in the solar system. Some of these differences can be explained by reference to the limitations of observing techniques, while others are quite incomprehensible. Overall, it would require admission that the solar system is something very special, and that the theories put forward to explain its formation do not apply universally.

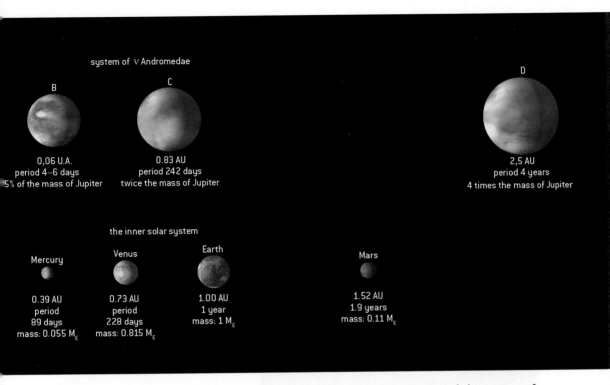

system of ν Andromedae

B

0,06 U.A.
period 4–6 days
5% of the mass of Jupiter

C

0.83 AU
period 242 days
twice the mass of Jupiter

D

2,5 AU
period 4 years
4 times the mass of Jupiter

the inner solar system

Mercury

0.39 AU
period
89 days
mass: 0.055 M$_E$

Venus

0.73 AU
period
228 days
mass: 0.815 M$_E$

Earth

1.00 AU
1 year
mass: 1 M$_E$

Mars

1.52 AU
1.9 years
mass: 0.11 M$_E$

On the same scale, a comparison of distances within the solar system and the system of the star υ Andromedae (the scale of masses differs). The planetary system of υ And, discovered in 1999, contains at least three planets, but there the resemblance ends. The planet (B) closest to the star is six times nearer to it than Mercury is to the Sun, but is a giant bigger than Saturn. The other two are even more massive, and have quite elliptical orbits.

Leaving aside the question of planets around pulsars, where do matters stand at present? Let us begin our answer by examining the least surprising findings. Among the two hundred exoplanets discovered by mid-2006, there are twenty-one multiple systems, and probably more if we take into account the difficulty of detecting planets smaller than Saturn. So, multi-planet systems like the Sun's are fairly common. The masses of these planets range from six Earth masses to about six Jupiters – which is not surprising, as anything smaller is undetectable. What *is* a surprise is that a good number of these exoJupiters are very close to their stars, and the orbital distance of many of them is less than that of Mercury around the Sun. Indeed, more than forty of them are less than 0.1 AU (Astronomical Units) from their stars, giving them, according to Kepler's third law, periods of the order of just a few days. Temperatures on these planets must be typically about 1,000° C – thus their common appellation: 'hot Jupiters'. They can have little in common, though, with the original Jupiter, 5.2 AU distant from the Sun!

Yet more bizarre are the orbits of hot Jupiters. They are almost circular, while some exoplanets at greater distances, with periods measured in years, follow markedly elliptical paths. These are far more elliptical than the orbit of any planet of the solar system.

This poses a real challenge for researchers accustomed to working with a theory of the solar system that explains why orbits there are near-circular, and why gas giants are far from the Sun while terrestrial planets orbit are close in. The real difficulty lies in understanding how giants can form close to stars. An obvious solution is to imagine that they could form far out within their systems, and then migrate inwards. Theoretically this is indeed possible, but two questions then arise. What causes the migration to cease, preventing the hot Jupiters from falling into their stars? And why, in our solar system, have Jupiter and Saturn remained where they are?

The solar system does seem to be quite an exceptional place. Will we one day find an exoEarth? Just finding a planet of terrestrial mass is not enough: it must not be too near (like Mercury or Venus) or too far (like Mars) from its star. If other planetary systems are so different from our own, we may not find it easy to discover a twin of the Earth!

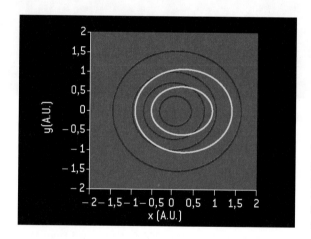

The system of the star **HD 74156**, discovered in 2004, comprises two planets comparable to Jupiter in mass. Their orbits (shown in yellow) are very elliptical, compared with orbits in the solar system (in red).

Solar system	Mass (M_E)	Semimajor axis (AU)	e^*
Mercury	0.055	0.387	0.206
Venus	0.949	0.723	0.007
Earth	1	1	0.017
Mars	0.107	1.52	0.093

HD 74156	Mass (M_E)	Semimajor axis (AU)	e^*
b	0.88	0.73	0.54
c	1.63	1.16	0.41

$^*e$ – orbital eccentricity; for a circular orbit, $e = 0$

Measuring exoplanets

The solar system is unlike any other known planetary system, but nevertheless is used as a standard, with Jupiter and the Earth as reference planets. Masses of exoplanets are measured in Jupiter masses (M_J), although those of some smaller bodies are expressed in Earth masses (M_E). Orbital radii are logically given in Jupiter radii (R_J), and distances from the central star in Astronomical Units. The orbital periods of exoplanets are given in days or years.

Mass of Jupiter: 1 M_J = 19 × 1,027 kg = 317.8 M_E

Mass of Saturn: 95.1 M_E

Mass of Earth: 1 M_E = 5.98 × 1,024 kg

1 Astronomical Unit (AU) = 149,600,000 km

Radius of Jupiter: 1 R_J = 142,800 km – 11.2 times the Earth's equatorial radius

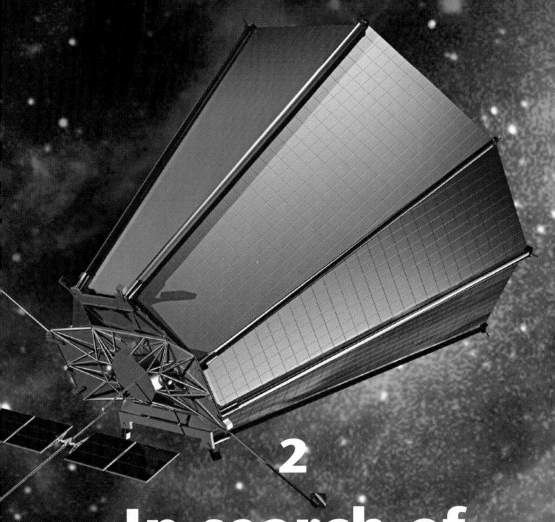

2
In search of exoplanets

The search for exoplanets extends back over several decades. Methods of detection are many and varied, including velocimetry, transits, gravitational microlensing and pulsar timing. They push the instruments involved to their limits. Actual images of exoplanets are as yet unfeasible, except in the case of some very special systems.

The American project Terrestrial Planet Finder (TPF) will consist of two complementary observatories. A visible-light coronagraph (seen here) is due to be launched after 2015, and an infrared interferometer will be launched after 2020.

2.1 THE PLURALITY OF INHABITED WORLDS

A question as old as astronomy itself

From Democritus and Epicurus to Aristotle and Seneca, Greek and then Roman philosophers often aired the question of the plurality of worlds. For Lucretius, 'there are, in other regions of space, Earths other than ours, different races of men, and different wild creatures.' Diogenes Laertius wrote: 'The Universe is infinite... of it, one part is a plenum, and the other a vacuum. He [Leucippus] also says that the elements, and the worlds which are derived from them, are infinite, and are dissolved again into them.'

More than a millennium later the debate took on a new dimension with the rise of the Copernican system, which taught that the planets revolve around the Sun. No longer was the Earth at the centre of everything. The question of the plurality of worlds was restated: if the Sun is but one star among all those strewn across the night sky, could not they also possess their retinues of planets? And might some of those distant worlds harbour life? At the beginning of the seventeenth century, Giordano Bruno was among the first to put forward the hypothesis: 'If the Universe be infinite, it must be that there exists a plurality of Suns... Around those Suns there may revolve worlds greater or smaller than our own.' This idea – considered heretical at the time – cost Bruno his life (although some of his other opinions and beliefs contributed to his demise). More than a century later, Christiaan Huygens was the first person to attempt direct observations of exoplanets; but he soon realised that this was impossible with the instruments then available. In the seventeenth century, philosophers and writers wondered about the possibility of inhabited worlds nearer to ours: might there be life elsewhere within the solar system? In the works of Fontenelle and Cyrano de Bergerac, the likelihood of living things upon Venus and the Moon was explored. These may have been rather vain speculations, but two centuries later the debate came once more to the fore with Giovanni Schiaparelli's announcement of the existence of *canali* on Mars – taken by some to be proof of an intelligent civilisation. Astronomical opinion remained divided on this question for a long time, until space probes settled it once and for all in the 1960s.

Cyrano de Bergerac on the Moon. Illustration from *Les Aventures Prodigieuses de Cyrano de Bergerac*, text and illustrations by Henriot (Epinal, c.1905).

'They leapt onto the rings of Saturn.' Illustration from *Micromégas*, by Voltaire. (Photogravure from 1867.)

However, spaceborne exploration of the Red Planet has, during the last two decades, firmly established that water once flowed there. The search for fossil life on Mars is still very much on the minds of astronomers.

The twentieth century saw further important steps in this field, with new observational methods allowing a dedicated search for the companions of nearby stars. Piet van de Kamp announced his discovery of a planet orbiting Barnard's Star, but methods available in the 1950s were not precise enough to allow confirmation of this.

Meanwhile, Stanley Miller's historic experiment in 1953 was a springboard for studies in prebiotic chemistry – the chemistry of the development of molecules necessary for the emergence of life. Echoing the work of the biochemist Oparin – who in 1924 had stated that life is the product of a long evolutionary process involving ever more complex organic chemistry – Miller was the first to synthesise amino acids from a gaseous mixture of water, methane and ammonia, into which electrical discharges were introduced. Many later experiments carried out under similar conditions have given rise to amino acids and nucleotides: the 'building blocks of life' involved in the synthesis of DNA.

Another surprise awaited researchers. Certain meteorites were found to contain traces of various prebiotic molecules and amino acids. Could life have been delivered to Earth on the asteroidal and cometary (meteoritic) material which bombarded it? And if this happened, might it not also have occurred on other distant planets? We now know that prebiotic molecules are abundant in comets and in the atmosphere of Saturn's satellite Titan, and are also found in the interstellar medium and in the environs of stars. Exobiology – the study of extraterrestrial life – is now undergoing a veritable revolution, driven by the discovery of exoplanets and by planetary exploration.

2.2 PROBLEMS WITH DIRECT IMAGING

2005: first images of exoplanets

In order to investigate the possibility of observing a planet orbiting a nearby star, let us construct an example. Imagine a 'pseudo-solar-system' in which a Jupiter-sized planet orbits a Sun-like star at a distance of 5 AU. The star is 5 parsecs (16 light-years) away from us. In the sky, the angle separating the planet from the

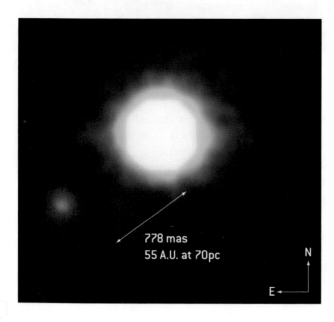

778 mas
55 A.U. at 70pc

N

E

The first image of an extrasolar planet, taken in 2004 with one of the 8.2-metre VLT telescopes (ESO) using the NACO adaptive optics system, which partially corrects for atmospheric distortion. The planet here appears in red, and at centre is the brown dwarf 2M 1207 (see opposite).

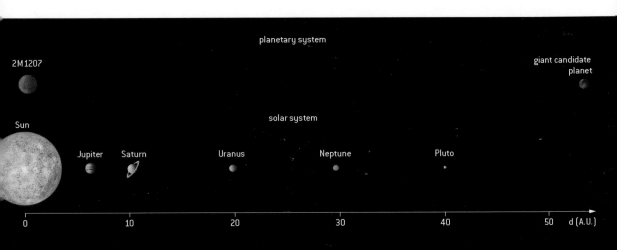

The system of the brown dwarf star 2M 1207 and its planetary companion. A team of French and American observers, studying the region around the very low-mass star 2M 1207, discovered a very faint object, apparently a companion to the star. This turned out to be a planet of about 5 Jupiter masses, orbiting at 55 AU from it (twice the distance of Neptune from the Sun). Later observations confirmed that the 'companion' was not a more distant star in the same direction, as the two objects moved together against the sky background. This, then, was the first time that a planet had been observed directly. Several favourable elements came into play in this case: the brown dwarf star is not very bright, and the planet is very big, with a contrast in brightness of only 100:1. The planet is a very long way from its star. Certainly, this is no exoEarth – not an environment where life is likely to flourish!

star would be 1 arcsecond (second of arc): 1/3,600 of a degree, equivalent to 1/1,800 the diameter of the full Moon. In theory, a sufficiently large telescope will separate and reveal two stars 1 arcsecond apart; but in the case of a star and its planet the contrast in brightness is far too great. Because the star is larger and hotter than the planet, it appears much brighter. The star's temperature would be of the order of 5,600 K, while that of our imaginary Jupiter would be 130 K. In visible light the star would seem about a billion times brighter than the planet! With a normal CCD camera it would not be possible to observe the planet, drowned in the light of its star. In the infrared, at a wavelength of 5–10 μm, the contrast is less unfavourable, but it would still be of the order of one million.

Direct imaging of planets with Earth-based instruments is therefore very difficult. However, in the case of giant exoplanets it will become possible during the years to come. There are two techniques which offer some hope. First, there is coronography, whereby the light from the central star is masked, just as the Sun can be masked during studies of its corona. Then there is nulling ('black fringe') interferometry in the infrared. This technique combines the signal from the star–planet system using a number of different telescopes working as an interferometer, in such a way that the light from the centre of the image (the star) is cancelled out. It seems likely, though, that the imaging of exoEarths will remain beyond the capacity of telescopes, and space missions are therefore being planned in order to achieve this ambitious goal. The American TPF (Terrestrial Planet Finder) and European Darwin projects are in process of development.

Other techniques have been studied for the indirect detection of planets orbiting nearby stars. The *astrometric* method is based on the principle of measuring the displacement of the central star in relation to the centre of gravity of the star–planet system. In the case of our imaginary solar system, this displacement would be only 0.001 arcsecond. The *Doppler–velocimetric* method involves measuring variations in the velocity of the star induced by the 'swaying' of the star around the centre of gravity. It is this second method which has led to the discovery of the majority of known exoplanets. Yet another technique is beginning to bear fruit: the observation of *transits* of exoplanets across the face of their stars. *Pulsar timing* and *gravitational microlensing* have also registered remarkable successes. In the following pages we shall examine some of the many detection methods in detail.

2.3 BARNARD'S STAR: A DISAPPOINTMENT

First attempts at detection, and first disappointments

In the nineteenth century, astronomers began to find evidence of the presence of smaller companions to stars when they detected slight periodic motions of those

Edward Emerson Barnard (1857–1923) was one of the greatest observers of his day. He discovered the fifth moon of Jupiter, Amalthea, in 1892, and is also remembered for the red dwarf star which bears his name.

Piet van de Kamp (1901–1995). In 1964 van de Kamp announced his discovery of a planet orbiting Barnard's Star, but in 1973 it was demonstrated that the star has no companion.

The Sproul Observatory at Swarthmore College, near Philadelphia. Here the Dutch astronomer Piet van de Kamp carried out studies of the motion and oscillations of Barnard's Star.

stars in relation to other nearby stars. The German astronomer–mathematician Friedrich Bessel was the first to detect such a body in this way – in this case the companion of Sirius. Sirius's partner, an extremely dense white dwarf star, was massive enough to affect the motion of its primary (Sirius A) without necessarily being detected visually.

A century later, improved techniques allowed astronomers to search for even smaller companions: for example, 'brown dwarf' stars or giant planets. By the 1940s the search for giant exoplanets was beginning. Patience was the key. Remember that within our solar system the periods of revolution of the giant planets around the Sun vary from 12 years in the case of Jupiter to 164 years in the case of Neptune. After several reports (later unconfirmed) of companions detected in the vicinities of the stars 61 Cygni and 70 Ophiuchi, a serious candidate took the stage when the Dutch astronomer Piet van de Kamp announced the discovery of a planet orbiting Barnard's Star. This star is a probable target for exoplanetary research, as it is the fourth closest star to the Sun. In 1964, on the basis of photographs taken over a twenty-year period, van de Kamp described the characteristics of the planet: its mass was 1.6 times that of Jupiter, and the period of its elliptical orbit was 24 years. As more data were produced, van de Kamp was able to refine his description, and in 1982 he concluded that there were two companions of Barnard's Star, with masses less than Jupiter's, and with periods of 12 and 20 years.

However, this spectacular result did not withstand the test of time. In 1973 a study revealed systematic errors in the performance of van de Kamp's Sproul Observatory telescope, which might have been responsible for the observed oscillations. In the same year the American astronomer George Gatewood published an independent study on the motion of Barnard's Star, and concluded that it had no companion. 'Barnard's Planet' may have had its day, but van de Kamp certainly counts as a pioneer in the field of research which is in outburst. The current avalanche of discoveries since planets were detected around a pulsar in 1992, has seen the first successes of the velocimetric method in 1995 and, in 1999, the first detection of an exoplanet by transit observations.

2.4 PLANETS AROUND PULSARS

Planets are found near dying stars

In recent times, astrometrists have sought exoplanets orbiting Sun-like stars, in the hope that they might one day find a world capable of supporting life. In the 1970s, however, another kind of star exercised their minds: pulsars (pulsating stars) had recently been discovered by radio astronomers. These objects represent the final stage in the evolution of some massive stars that have exhausted their supplies of hydrogen and helium. Heavier elements are later synthesised, and they become supergiant stars, and explode as supernovae, ejecting most of their material into space. The central remnant collapses upon itself to form an extremely dense neutron star – its mass comparable to that of the Sun, but packed into a volume about 20 km in diameter! Neutron stars rotate very rapidly, with periods of the order of 1 second. Pulsars exhibit very strong (synchrotron) radio emissions, keeping time with their rotation. Radio astronomers detect these highly regular pulses: the signature of pulsars.

Detecting exoplanets around pulsars

Pulsar timing is the most sensitive method for detecting pulsar planets. It involves the study of variations in the pulsar's period, expressed in milliseconds. These variations are of the order of 1.2 $[M_p]$ $[P]^{2/3}$, where P is the period of revolution of the planet in years and M_p is the mass of the planet in Earth masses, assuming a circular orbit. Since this method is extremely precise, tiny variations in the period can be detected: for example, near PSR 1257+12 a companion of 0.020 Earth masses (1.6 times the mass of the Moon) has been found. By this method, even 'exoMoons' are detectable. Note that the sensitivity of this method does not explicitly depend upon the distance of the pulsar. However, difficulties remain, as the signals from more distant pulsars are weak, and timings of them are less accurate.

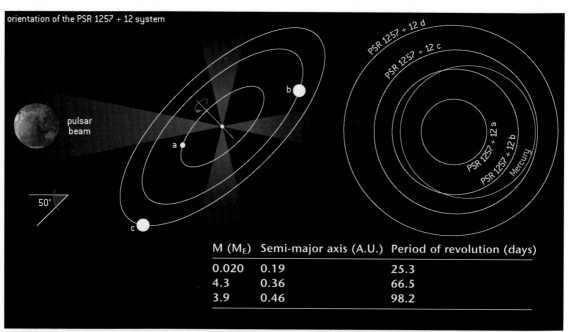

orientation of the PSR 1257 + 12 system

pulsar beam

50°

M (M_E)	Semi-major axis (A.U.)	Period of revolution (days)
0.020	0.19	25.3
4.3	0.36	66.5
3.9	0.46	98.2

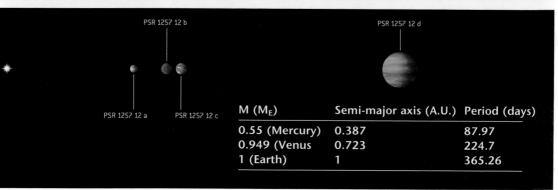

PSR 1257 12 b

PSR 1257 12 d

PSR 1257 12 a

PSR 1257 12 c

M (M_E)	Semi-major axis (A.U.)	Period (days)
0.55 (Mercury)	0.387	87.97
0.949 (Venus	0.723	224.7
1 (Earth)	1	365.26

The planetary system of PSR 1257+12. Above left: at least three planets orbit PSR 1257+12. The masses of the two largest (b and c) are 4.3 and 3.9 times that of Earth, while the smallest (a) is a little more massive than the Moon. Above right, and below: planets a, b and c have near-circular orbits very close to the star, at distances smaller than that of Mercury from the Sun. If the central star were not a pulsar, this would be the exoplanetary system which most resembles the solar system! Another pulsar with planets is PSR 1620–26. Its system – with one planet of 2.5 Jupiter masses orbiting at 40 AU – is quite unlike that of PSR 1257+12.

Very accurate measurements of radio signals from pulsars provided a new method in the search for possible exoplanets, the presence of which might be revealed by variations in the periods of these stars. Several 'detections' were announced, but it later transpired that they were caused by the pulsars' internal instabilities. Then, in 1991, British astronomer Andrew Lyne published his discovery of a Uranus-sized companion to PSR 1829–10 – a pulsar 30,000 light-years away. The orbital period of the planet was calculated to be 6 months. But disappointment continued to stalk the astronomers. The observed effect was not due to the presence of a planet, but to an inaccurate 'too circular' correction for the elliptical motion of the Earth around the Sun. Once this was corrected, the irregular periodicity, and the planet, disappeared.

Finally, as Lyne came to terms with the disappointment of a negative result, the long search for a 'pulsar planet' came to an end. In 1992, Polish astronomer Aleksander Wolszczan announced the discovery of two planets orbiting pulsar PSR 1257+12. This is a very different pulsar from that previously mentioned. Its period is an incredibly short 0.0062 seconds, placing it in the category of millisecond pulsars. These have life stories which are not at all the same as those of classical pulsars. Millisecond pulsars are much older, having existed for hundreds of millions of years. Their peculiar nature can be explained thus. They are the end product of a pair of stars – one normal and the other a neutron star. As the normal star ages it dilates, transferring momentum to the neutron star and spinning it up. How can such an object possess planets? This must be a result of the accretion of stellar matter into a disk around the neutron star, and it is within this disk that planets are later formed. Such worlds are doubtless very different from the kind of planets with which we are familiar.

It was later discovered that PSR 1257+12 has three planets, with masses between 0.02 and 100 times the mass of the Earth. Apart from a planet 2.5 times the mass of Jupiter, companion to pulsar PSR 1620–26 and discovered in 1999, no other 'pulsar planets' have yet been detected near millisecond or standard pulsars. This suggests that they are rare objects – especially since detection methods are very sensitive.

2.5 THE KEY TO SUCCESS: VELOCIMETRY

The Doppler effect to the rescue

Let us return to the exoplanets which interest us most: those orbiting stars in the vicinity of the Sun. The negative conclusion of the work on Barnard's Star dashed the hopes of classical astronomy that sufficiently accurate measurement of tiny oscillations in central stars might be possible. However, in the 1990s another method came into its own: velocimetry.

How does this technique work? The procedure is based on measuring very accurately the radial velocity of the star (its velocity *vis-à-vis* the observer) as a function of time, in order to identify any periodic movement of the star around the common centre of gravity of the star and its planet. The movement of the companion cannot be measured, as it is not bright enough to be detectable. The method can be used if the orbit of the planet to be detected is not perpendicular

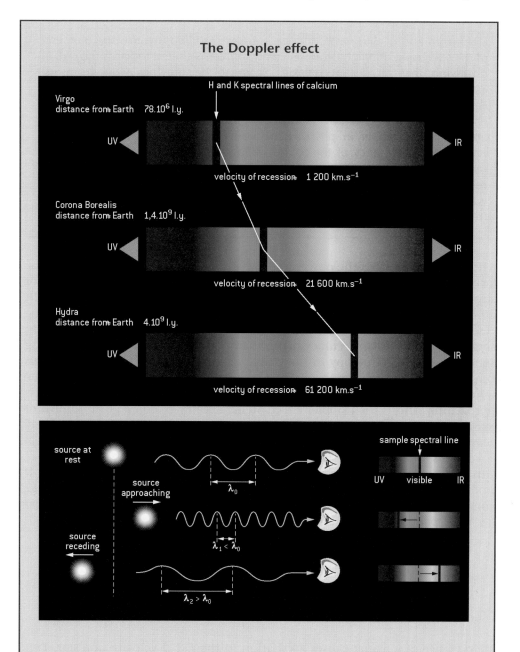

The principle of the Doppler effect. If an object moves relative to the observer, the spectral lines are shifted from their initial rest position: towards the red in the case of a receding object, and towards the blue in the case of an approaching object. By using large numbers of known lines, accuracy to within a few m/s can be achieved.

Light from approaching astronomical sources is shifted towards the blue end of the spectrum, and light from receding sources is shifted towards the red. In observing that all distant galaxies seem to be moving away from us, astronomers use this as evidence of the expansion of the Universe. Here on Earth it is easy to experience the Doppler effect by listening to the sound emitted by a passing vehicle such as a motorcycle or ambulance. The sound appears to be higher in pitch as the vehicle approaches, and lower as it recedes. The frequency of the sound waves is therefore lower as the vehicle recedes, implying that the wavelength is greater (frequency has an inverse relationship to wavelength). Light-waves behave in the same way: the wavelengths of the light from a receding object appears lengthened (redshifted), compared with that from an object at rest.

to the line of sight from the observer to the star. If it is, the radial velocity of the star is zero. It is worth mentioning that this infrequent case is favourable for astrometry. It remains only to measure the radial velocity of the star with sufficient accuracy. The rate of the periodic movement of the Sun due to the presence of Jupiter is 12.5 m/s. If we wish to find an exoJupiter we will need to measure velocities to an accuracy better than a few metres per second, requiring high-resolution spectroscopy in the visible region. We know that if a luminous body is moving relative to an observer, the lines in its spectrum appear to be shifted towards the red end if it is receding, and towards the blue end if it is approaching. This is the Doppler effect. Stellar spectra exhibit a great number of well-known spectral lines, and their Doppler displacement can be accurately measured. If many lines are studied, accuracy can be increased to reveal movements of a few metres per second only.

In addition to accuracy we must include consistency over a long period. If another 'Jupiter' is to be found in some distant system, the search will take 12 years as the planet pursues its full orbit, causing one complete oscillation of that star. Only patient astronomers need apply.

2.6 THE INS AND OUTS OF VELOCIMETRIC DETECTION

A sensitive method, but with a few drawbacks

The presence of a planet causes a variation in the radial velocity of a star. This depends on several criteria: the mass of the star, the mass of the planet, the period of the planet's orbit, and the orbital eccentricity. An approximation might begin with the simplest case: that of a planet in a circular orbit. Here, variations in velocity are sinusoidal, with a period matching the orbital period P of the planet. Their amplitude K in m/s is expressed by

$$K = 28.4 \, P^{-1/3} \, (M_p \sin i) \, M_*^{-2/3}$$

using 'natural' units: the period of revolution P in years, the mass of the planet in Jupiter masses, and the mass of the star in solar masses. If Jupiter were 1 AU from the Sun it would cause a variation in radial velocity of our star of 28.4 m/s, whereas in the case of the Earth the value is only 0.1 m/s! It should be noted that it is not the mass of the planet which is directly involved, but the quantity $M_p \sin i$, where the angle i defines the inclination of the orbit of the planet as seen from

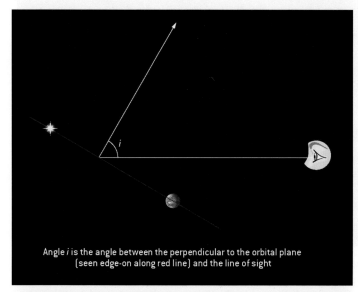

Angle i is the angle between the perpendicular to the orbital plane
(seen edge-on along red line) and the line of sight

Definition of angle i, defining how an observer on Earth sees the orbit of the planet observed. When i is 90° the orbit is seen edge-on, and when it is 0° it is seen face-on.

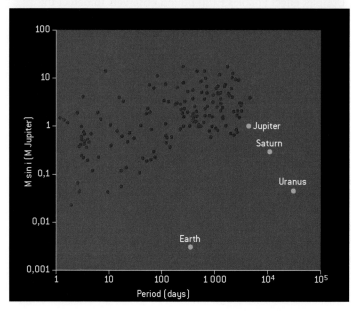

The relationship between minimum mass/period of revolution of extrasolar planets (in red) and some solar system planets (in blue). The velocimetric method is incapable of detecting giant planets at distances from their stars comparable to those of terrestrial planets (except in the case of Jupiter). It is therefore no surprise that astronomers have not yet detected a system similar to our own.

The principle of the velocimetric method of detecting extrasolar planets. (a) The planet orbits its star, or more precisely, they both orbit about G, the barycentre of the star–planet pair. (b) The Doppler effect enables Earth-bound observers to detect the star's motion as induced by a large enough planet or a collection of planets. The alternating shifts towards the red and the blue, characteristic of stars being perturbed by their planet(s), may be graphically represented by a sinusoidal curve.

Earth (see figure on p. 25). If this angle is 90°, the orbit is seen edge-on. In this case, sin i is 1, and the variation in velocity is maximised. If the angle is 0°, then the orbit is seen face-on, and sin i is 0, causing no variation in radial velocity. In the usual case, where the orbit is presented at an intermediate inclination, angle i is between 0° and 90°, and M_p sin i lies between 0 and the actual value of the mass of the planet. With no other source of information than the measurement of radial velocities, neither the inclination of the orbit nor the value of sin i are known. This is the great drawback of this method: it can be used to determine not the mass of the planet, but the *minimum value* of that mass. The parameters obtained are therefore the minimum mass, the period (directly linked with the semimajor axis a of the orbit, according to Kepler's third law), and the eccentricity of the orbit if not circular.

The method is not without its caveats. The more massive the planet, the more marked is the perturbation of the radial velocity. The same is true in the case of short periods. The velocimetric method is therefore well suited to the study of massive, short-period planets, close to their stars; in short, 'hot Jupiters'. Another factor working against the discovery of 'outer' planets is that, for the unambiguous detection of such a planet, the duration of the study should be at least equal to its orbital period.

Remember that the theoretical limit of 1 m/s is difficult to achieve, since some of the stars in question are variable, as is the case with the Sun. These variations introduce inevitable 'noise' into the observations. Moreover, a star's intrinsic periodic variations also have to be taken into account (the Sun, for example, has a cycle of activity of ∼ 11 years), as such variations might be misinterpreted as being caused by planets. As with the detection of pulsar planets, the method is not dependent upon the distances of stars, but in reality it is difficult to achieve accurate measurements of radial velocity for very distant (and therefore fainter) stars.

2.7 51 PEGASI B: THE FIRST DISCOVERY

1995: first results from velocimetry

In the early 1990s, several teams of astronomers embarked upon systematic searches for low-mass companions of nearby Sun-like stars. Their targets were not only exoplanets, but also brown dwarfs – embryonic stars that have insufficient mass (less than 0.08 times the mass of the Sun) to trigger the fusion reaction to 'burn' hydrogen into helium. This mass is equivalent to about 80 times the mass of Jupiter, and so brown dwarfs are *a priori* less difficult to find than exoplanets.

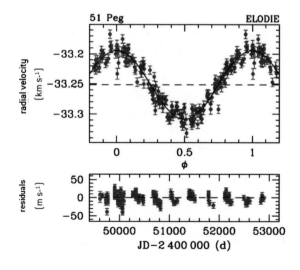

The velocimetric curve for 51 Pegasi, derived by Swiss astronomers Michel Mayor and Didier Queloz in 1995, using the high-resolution ELODIE spectrometer at the Haute-Provence Observatory. (Upper) Radial velocity as a function of phase, adjusted to a sinusoidal curve of 4.2 days. (Lower) The departure from the curve as a function of time.

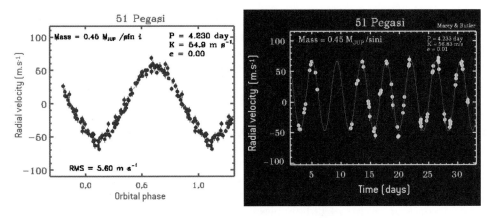

The velocimetric curve for 51 Pegasi derived by American astronomers Geoffrey Marcy and Paul Butler in 1997, at the Lick Observatory in California. (left) Radial velocity as a function of orbital phase. (right) Velocity as a function of time (over one month beginning 12 October 1995).

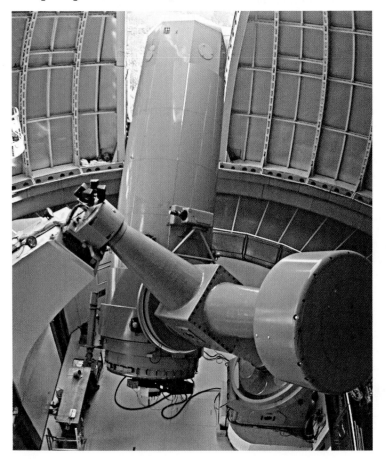

The 193-cm telescope at the Haute-Provence Observatory. The telescope was installed in 1958, and the ELODIE high-resolution spectrograph, built at the observatory, began its work in mid-1993.

The three teams deeply involved in the hunt were led by Bruce Campbell and Gordon Walker in Canada, Geoffrey Marcy and Paul Butler in the USA, and Michel Mayor and Didier Queloz at the Geneva Observatory. The Canadians in particular carried out much pioneer work in the development of the instrumentation involved in this research, and in 1994 they announced a find which was not confirmed by later investigations. In 1994 Marcy stated that after two years of research a review of a third of his target stars had yielded negative results. Meanwhile the Swiss team had begun a systematic study using a new high-resolution spectrometer, working with the 1.93-metre telescope of the Haute-Provence Observatory in southern France. In autumn 1995 they announced the first discovery of an exoplanet orbiting a 7 billion-year-old Sun-like star: 51 Pegasi. Its companion appeared to be at least half as massive as Jupiter, with an orbital period of only 4.2 days. So, here was a planet orbiting a star at a distance of only 0.05 AU (a twentieth of the distance of the Earth from the Sun)! In the weeks which followed, Marcy was able to confirm the discovery and announce the detection of two other exoplanets – one orbiting the star 47 Ursae Majoris and the other orbiting 70 Virginis, and both pursuing very elliptical orbits. With masses of at least 2.6 and 7 times that of Jupiter, these planets were at mean distances of 0.5 and 2 AU from their respective stars.

It immediately became clear that astronomers were dealing with distant stellar systems very different from our own; and this evidence was confirmed by later discoveries.

Naming exoplanets

The names of exoplanets are hardly poetic. Let us take the star HD 38529 as an example. The first exoplanet discovered orbiting this star is designated HD 38529 b, the next HD 38529 c, and so on. HD 38529 itself derives its name from its number in the star catalogue of Henry Draper. Most of the stars observed by the velocimetric method will be nearby stars such as HD 38529, already listed in various catalogues, or ε Eridani, the fifth brightest star (in apparent magnitude) in the constellation of Eridanus. 51 Pegasi is number 51 in Pegasus, according to the system of Flamsteed, who numbered stars within constellations in order of right ascension (from west to east). Some other stars are numbered according to their sky coordinates. In the case of PSR 1257+12, PSR signifies that it is a pulsar, and the numbers refer to its approximate right ascension (12 h 57 min) and declination (+12°). There are also planets with designations that reflect the names of the research programmes which detected them: for example, OGLE-TR-56 b was found by astronomers working on the Optical Gravitational Lensing Experiment, and TrES-1 was the first planet revealed by the Transatlantic Exoplanet Survey. We may rest assured, though, that some planets which will be studied in detail will receive more attractive names: HD 209458 b is already 'Osiris', according to those closely connected with it!

2.8 WHEN PLANETS CROSS THE DISKS OF STARS

The search for exoplanets using transits: the first successes

Velocimetry has been used to find more than 200 giant exoplanets in less than twelve years; yet it still carries the great disadvantage of providing only a lower limit for the mass of a planet, since the angle of inclination of the orbital plane of the planet *vis-à-vis* the observer is unknown. Neither can the method reveal any information about the physical nature of the planet; for example, its diameter.

There is, however, another method which can circumvent these problems, especially in combination with velocimetry: the transit method. This is used when the plane of the planet's orbit is observed edge-on (see diagram below). This configuration causes the planet to occasionally move across the face of its star, thereby hiding a small part of the star's surface. Similar transits are observed locally when, for example, Venus passes in front of the Sun. During the transit the luminous flux received from the star will be slightly reduced. In the case of a planet like Jupiter, with a diameter of about 0.1 that of its star, the reduction will be of the order of 1%. An Earth-sized planet, with a diameter 0.01 that of its star, will involve a reduction of 0.01%.

The first condition for the use of this method is, of course, that the planet should (from the observer's viewpoint) pass in front of its star; but the

The planetary transit method. When the planet passes in front of its star, the luminous flux received from the star will be reduced. In the case of a planet like Jupiter, the reduction (comparing the situation when the planet is in position 1 and position 3) will be of the order of 10^{-2}; and for an Earth-sized planet, of the order of 10^{-4}, assuming a solar-type star.

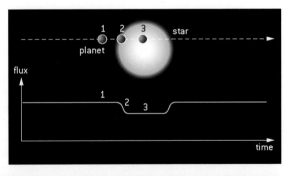

Transit of an exoplanet observed simultaneously by occultation and velocimetric methods.

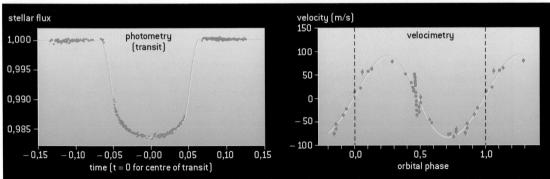

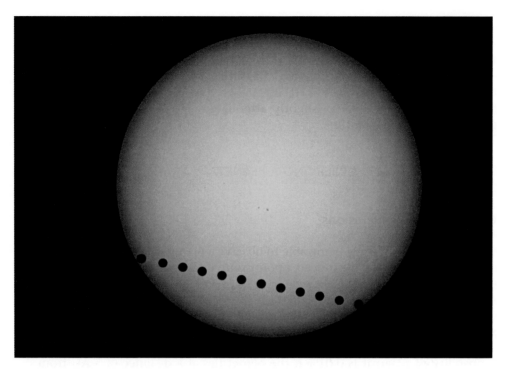

Transit of Venus across the face of the Sun on 8 June 2004, based on successive images taken at Waldenburg, in Germany – the first at 05.41 GMT and the last at 11.04 GMT. This was the first transit of Venus since 1882. Venus is about the same size as the Earth, and its orbit is closer to the Sun. A transit does not occur every time Venus passes this side of the Sun, since the orbits of the Earth and Venus are not in the same plane. These transits occur in pairs separated by 8 years, every 105.5 (105.5+8) and 121.5 (121.5+8) years.

probability of this occurring also depends on the size of the star, and is greater when dealing with a large star. The distance of the planet from the star is another factor, and the nearer it is, the greater is the chance that a transit will occur.

Astronomers systematically carried out photometric studies of all the stars known to have companions discovered by velocimetry, in the hope of detecting transits. In the case of HD 209458 b they struck lucky. Based on velocimetric data, the first transit of an exoplanet was observed by David Charbonneau and Timothy Brown in September 1999. They were able to determine not only the radius of the planet, but also its orbital inclination as seen from Earth, from which data they were able to calculate its exact mass. It is 0.69 times the mass of Jupiter, and its diameter is 1.5 times greater than Jupiter's diameter. This is therefore not a very dense exoplanet (0.3 g/cm^3) as confirmed by existing theoretical models. A transit of HD 209458 b was later observed, with even greater accuracy, by the Hubble Space Telescope.

Archive material from the astrometric satellite Hipparcos also confirmed the existence of HD 209458 b. Hipparcos was launched by the European Space Agency in 1989, and over many years it determined the positions of more than 100,000 stars, with a view to establishing a comprehensive and highly accurate catalogue of their proper motions and magnitudes. Its data revealed evidence of a transit of a planet across the star HD 209458 on five separate occasions, when the light from the star was very slightly attenuated. This unusual behaviour had passed unnoticed at the time.

2.9 OBSERVING A PLANETARY TRANSIT

Conditions and conclusions

What is the probability of being able to observe a planetary transit across a star at distance a? If the star were indeed a point source, the transit would occur only if the star, the planet and the observer were in exact alignment. However, since the star has a definite radius R_*, calculation shows that the probability of transit is expressed by R_*/a. Simple logic: the nearer a planet is to its star, the greater the possibility of a transit; and the larger the star, the greater the chance that the planet will be seen to pass across it. In the case of a star the size of the Sun, the probability of a transit is 0.01% if the planet lies at a distance of 5 AU from its star. For a planet at 1 AU, the probability is 0.5%, rising to a not insignificant 10% if the planet orbits at 0.05 AU.

	Probability of transit %	Duration of transit (hours)	Variation in flux %
Mercury	1.2	8	0.0012
Venus	0.64	11	0.0076
Earth	0.47	13	0.0084
Jupiter	0.09	30	1
Saturn	0.05	41	0.75

In the case of the solar system as seen from outside, the further the observer from the Sun, the less probable it becomes that a transit will be observable. The duration of the transit increases with the distance of the planet from the star. The variation in flux depends on the size of the planet, and is of course greatest in the case of Jupiter.

	Probability of transit %	Duration of transit (hours)	Variation in flux %
51 Pegasi b	9.1	3	1
HD 209458 b	10.8	3	1.6

The configuration of hot Jupiters – large planets close to their stars – favours the visibility of transits. The probability that the planet, as seen from Earth, will cross the face of the star attains 10%.

Space missions employing the transit method

The detection of exoEarths using ground-based techniques seems to be beyond our capabilities, since photometric accuracy of the order of 10^{-4} is required. For this reason, space missions ensuring the necessary photometric stability – the French COROT and the American Kepler missions – have been or will be launched in the near future. These two projects might also be able to detect possible satellites or rings of exoplanets by accurate measurement of light-curves during transits.

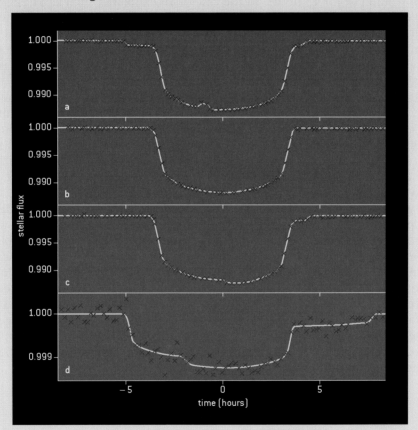

Deformation of transit light-curves due to the presence of a satellite. Examples of light-curves calculated for different configurations of exoplanets and their satellites. The first three curves correspond to a Jupiter-sized planet with a satellite of period 0.5 days (a) or 1.5 days (c), or without a satellite (b). Case (d) corresponds to a smaller exoplanet (2.5 Earth radii) with a slightly smaller satellite. (After Sartoretti and Schneider, *A. & A. Suppl.*, **134**, 553 (1999).)

The diminution in brightness due to the interposition of the planet is easy to determine. It is simply the ratio of the apparent surfaces of the planet and the star: $\Delta F = (R_p/R_*)^2$, where R_p is the radius of the planet. Consequently, if we are able to estimate the radius of the star we can deduce the radius of the planet. The duration of the transit depends firstly upon the period of revolution of the planet around the star (the further away the planet is from the star, the longer it will take to pass across its face), and secondly, upon the inclination of the orbit. For planets at different distances, the transit may last for periods ranging from hours to several days. Seen from outside the solar system, a transit of the Earth across the Sun would last for 13 hours. In the case of an exoplanet such as 51 Pegasi b, only 0.05 AU from its star, the duration of a transit would be just 3 hours, assuming that the star is the size of the Sun. Another factor worth noting is that it is impossible to observe an object 24 hours a day from the rotating Earth, because of the alternation of night and day – unless you are based in the polar regions! More problems to solve. . .

In the final analysis, from transit observations we can deduce the radius of the planet, its period, and the inclination of its orbit – but not its mass. Complementary, velocimetric observations are needed, enabling the minimum mass of the planet to be determined, and, since the orbital inclination is now known, the value of the mass of the planet. With mass, radius and distance from its star all known, some physical studies can be undertaken; but it must be said that divining the nature of exoplanets is something of a treasure hunt!

The game is worth the candle, but the difficulties are many. In order to detect an exoplanet by the transit method, the luminous flux from the star has to be metered very accurately over an extended period, in the hope of being able to observe regular fluctuations as the star dims slightly for a certain repeated amount of time. Accuracy to at least 1% in the measurement of the flux is necessary for the detection of giant planets. To increase the chances of success, fields containing millions of stars are studied, entailing the analysis of mountains of data. From this process, hundreds of candidate stars emerge, to be observed by other methods. The vast majority of these candidates will be revealed as variable stars or eclipsing binaries; but a mere handful will be labelled 'exoplanet'. So far, nine have been found or observed by this method.

2.10 THE GRAVITATIONAL MICROLENSING METHOD

When a star amplifies the light from another star: enter general relativity

With astrometric, velocimetric and transit methods already in use, a new and original technique for finding exoplanets emerged. It is based upon an application of Einstein's theory of general relativity, according to which light-rays are bent in the vicinity of very massive objects. This effect was confirmed during the total solar eclipse of 1919, when astronomers accurately measured

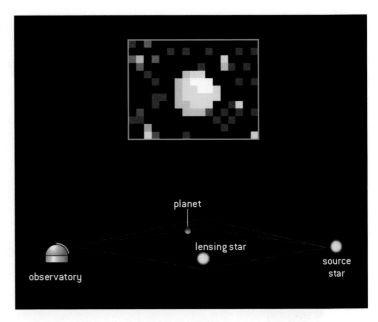

The principle of gravitational microlensing. The observer on Earth sees the source (distant) star when the lensing (nearer) star passes across the centre of the image. The framed image shows what may be seen with a ground-based telescope. General relativity tells us that light-rays from distant objects are bent as they pass near (very massive) closer objects. The gravitational field of a star or a galaxy may therefore act as a gravitational lens, amplifying the light from some background objects.

planet

lensing star

source star

observatory

positions of distant stars near the Sun's limb, by which the predicted displacement was observed. We can therefore liken a star's or a galaxy's gravitational field to a 'gravitational lens', able to bend light. Such lenses are able to amplify the light from stars behind them. If a star passes exactly in front of a more distant star, the flux from the latter is, at the exact moment of 'eclipse', much amplified by the gravitational lens as it causes the rays to converge in the direction of the observer.

This effect has been used in the systematic search for very faint stars in our galaxy: the famous brown dwarfs, for example. The method somewhat resembles the transit method. A careful survey of a star-field containing thousands or even millions of stars is undertaken; but instead of watching for a decrease in brightness, astronomers seek, on the contrary, temporary increases in brightness as a star, too faint to be seen otherwise, crosses the line of sight. This is the gravitational microlens effect. What will be seen is a light-curve exhibiting a maximum flux for a period which may last from days to months.

The search for exoplanets by this method is a by-product of the search for low-mass stars. If the microlensing star has a planet, the generated light-curve will show an anomaly as the planet passes in front of the more distant star. It is a productive method, as four planets have already been detected in this way – one in 2003, and the others (one of which is the smallest exoplanet found so far, with a mass six times that of Earth) in 2005. The problem with this method, however, is that the phenomenon cannot be predicted, as it requires a particular configuration to occur; and it is not reproducible, as the passage of the star and its exoplanet in front of the distant object is a singular event. The microlensing method therefore involves the systematic observation of an

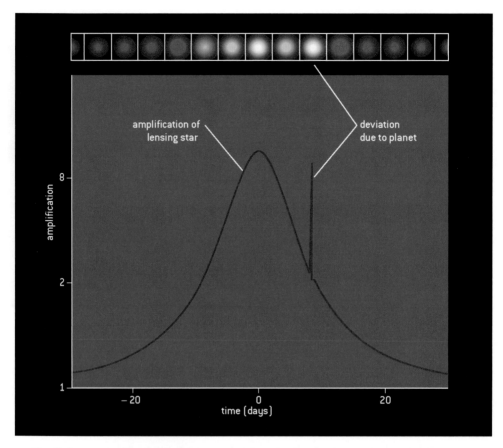

The light-curve of a distant star, obtained using the gravitational microlensing method. The passage of an Earth-sized planet across the face of a lensing star may disturb the light-curve due to the lensing star – briefly but spectacularly. The gravitational microlensing method may be used to detect Earth-sized planets from the ground. (After a theoretical calculation by Bennett and Rhie, *Ap. J.*, **472**, 660 (1996).)

especially dense field of stars, as is found, for example, towards the centre of our galaxy. There are several studies in progress, including PLANET (Probing Lensing Anomalies NETwork), OGLE (Optical Gravitational Lensing Experiment), and MOA (Microlensing Observations in Astrophysics). To date, few brown dwarfs have been discovered; but perhaps the search for exoplanets may prove more successful. It must also be admitted that although OGLE has detected several exoplanets, most of them have been revealed by the transit method, and not via microlenses!

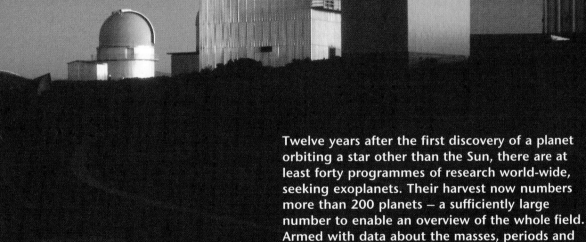

3
Twelve years of discovery

Twelve years after the first discovery of a planet orbiting a star other than the Sun, there are at least forty programmes of research world-wide, seeking exoplanets. Their harvest now numbers more than 200 planets – a sufficiently large number to enable an overview of the whole field. Armed with data about the masses, periods and orbital eccentricities of giant exoplanets, astronomers can now try to understand how they form and evolve.

The La Silla Observatory, Chile. Several telescopes participating in the search for exoplanets operate at this site – among them the 3.6-metre telescope and the Swiss 1.2-metre telescope.

3.1 A VERY SELECTIVE METHOD OF DISCOVERY

The radial velocity method shows a marked preference for short-period, massive planets

What do we know about other planetary systems? Very little, if we consider individually each of the planets so far discovered. The characteristics deduced using the radial velocities method (the only method that has so far yielded a good number of results) involve three parameters: the period of revolution of the planet (directly linked with the semimajor axis of its orbit, according to Kepler's Law); the ellipticity of the orbit; and the planet's mass, if the inclination of the orbit is known, involving the factor sin i. Under the right circumstances, the number of planets involved in the system may also be deduced. These parameters are decided on the basis of observed variations in velocity, and depend on the mass of the star involved, estimated using spectral type/luminosity models. As for such a basic parameter as the radius of the planet, this is only known in the case of the objects observed by the transit method. With the available evidence we cannot really step into the realms of speculation about the individual nature of such planets or their composition or climate; but with more than 200 already known, statistical studies become possible, and the global properties of the population of exoplanets can be discussed.

As with any statistical study, we must take into account factors which might introduce bias into our interpretation. Not all planets are detectable. Far from it! Certainly, all observable solar-type stars come under scrutiny – the description 'observable' meaning that they are bright enough and stable enough to be studied using the radial velocity method. In practice this limits the search to stars nearer than 50 parsecs. Within this sample, a planet with a period longer than the time allotted for the search programme may not be detected. Currently, it is just possible to detect a planet of 1 M_J if it is, like Jupiter, at a distance of 5 AU from its star. This would give it an orbital period of 12 years, which is about the duration of searches, the earliest dating from 1995.

As far as planetary masses are concerned, the amplitude K of the perturbations of the star's velocity is directly proportional to the minimum mass of the planet, M_p sin i. Unsurprisingly, high-mass planets are the easiest to detect. Remember that the detection limit of the best instruments is around a few m/s, the goal being to attain 1 m/s. Detection of an 'Earth' would require better than 0.1 m/s. K also depends to a lesser extent on the planet's period of revolution, linked to its mean distance from the star (longer periods being the hardest to detect) and the eccentricity of the orbit. The cumulative effects of these parameters are therefore quite subtle. Lastly, it must be pointed out that the theoretical accuracy of 1 m/s could be achieved only in the case of a star with an absolutely calm atmosphere, without spots or prominences. In reality this is improbable, and care must be taken not to confuse the effects of stellar activity with signs of the presence of a planet.

There is one more adjustment to consider – by no means the least important. What we know is the minimum mass of the exoplanet, since the method underestimates true masses by a factor 1/ sin *i*. This can be worked in only statistically, and on average the true mass of the planet will be 1.57 times greater than that arrived at through observation.

Limits of detection using the radial velocity method

The radial velocity method is sensitive to perturbations induced in the observed velocity of a star, but in practice the limit of detection of current instruments is of the order of a few m/s. For example, the ELODIE spectrograph attached to the 193-cm telescope at the Haute-Provence Observatory is accurate to 8 m/s. CORALIE, on the Swiss 1-metre telescope at La Silla, in Chile, is accurate to 5 m/s. The HARPS instrument on the 3.6-m telescope at La Silla reaches a limit of about 1 m/s.

There is also the question of the length of time taken to observe the star. A Jupiter-mass planet would induce an effect of 2.8 m/s if it orbited at 100 AU from its star; but it would not be detected, because its orbital period would be 1,000 years. Radial velocity measurements have been undertaken system-atically since 1995 – which limits the detection of any planet with an orbital period shorter than 12 years, whatever its mass.

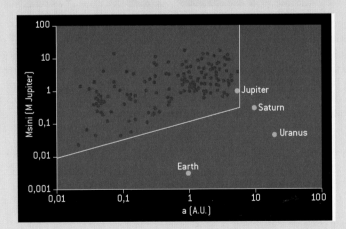

The minimum mass of planets M_p sin *i* as a function of the semimajor axis of their orbits, *a*. The amplitude *K* (in m/s) of the perturbations in the velocity of the star in question is directly proportional to the minimum mass of the planet orbiting it, but it also depends on the planet's orbital period *P* and the eccentricity *e* of the orbit, according to the equation:

$$K = 28.4 \, (P/1 \text{ year})^{-1/3} \, (M_p \sin i/M_J) \, (M_* / M_0)^{-2/3} \, (1 - e^2)^{-1/2}.$$

The High Accuracy Radial Velocity Planet Searcher (HARPS) spectrograph. With this instrument, mounted on the 3.6-metre ESO telescope, Michel Mayor's team is reaching an accuracy of 1 m/s in the measurement of radial velocities. Components of great accuracy and unusual size were needed in the construction of HARPS. This photograph – taken during laboratory testing, with the vacuum chamber of the spectrograph open – shows the large optical system which disperses starlight in such a way that its spectrum can be accurately measured. The dimensions of this system are 20 × 80 cm.

3.2 EXOPLANETS: A HEAVYWEIGHT FAMILY?

Plenty of Jupiters, but few 'super-Jupiters'

Bearing in mind all the caveats mentioned, what can we now say about the large family of exoplanets? Let us first look at their minimum masses, which range from 0.02 M_J to 17.5 M_J. The lower figure represents merely the limit of detection in the most favourable case. The histogram of masses shows a marked preponderance of masses below 2 M_J, and half the planets detected are of mass less than 1.6 M_J. Many low-mass planets must have 'fallen through the net', as

The birth of the solar system. According to the standard theory of the formation of the solar system, the Sun formed at the heart of an immense disk of gas (hydrogen and helium), dust and ice. The disk collapsed under its own gravity, but as it was rotating it settled into a disk with the protoSun at its centre. Within the disk, embryonic planets (planetesimals) grew as they gathered up dust particles. In the central regions, ice particles were melted by solar radiation, and dust agglomerated to form rocky planets. In the outer regions there was sufficient material (dust, and especially ice) to form very large planetesimals, which became massive enough to draw the surrounding gas in upon themselves. The giant planets were born. This scenario explains the spatial separation between the terrestrial planets and the giants. It also explains why the orbits of the planets in the solar system are more or less in the same plane – that of the initial disk – and why these orbits are almost circular. Any planetesimals with orbits too elliptical would be destroyed in collisions with others. Unfortunately, the properties of most exoplanetary systems do not fit this scenario.

they are much more difficult to detect than their larger neighbours, and there must be an even greater preponderance of small fry. This distribution – with its obvious preference for 'low' masses – is therefore quite a firm result, independent of random selection effects. Here we see a reflection of our own solar system: of its eight planets, the small ones are as numerous as the large.

The form of the distribution is not in itself remarkable, and the 'standard' theory of the formation of the solar system explains it well enough. If planets form by accretion, from protoplanets or planetesimals within a protoplanetary disk, logic suggests that more small objects than large ones will result, since more massive planets require more material. However, a nagging doubt remains. How can we explain the presence of really large planets of about 8–10 M_J? Such bodies would need tens of millions or even hundreds of millions of years to form; but all observational evidence indicates that after a few million years (10 million at most), young stars have lost their protoplanetary disks. How does this happen? All stars lose matter in the form of a continuous 'wind' of particles and eruptions. The Sun is no exception, and the solar wind is responsible for magnetic storms and polar aurorae on Earth. Young stars, however, emit fierce stellar winds,

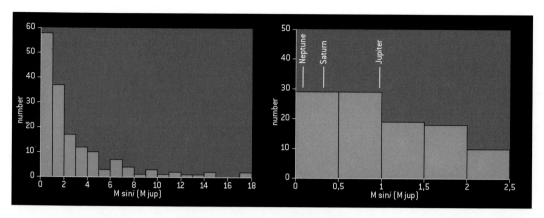

The masses of exoplanets. The observed distribution of the 161 exoplanets known by July 2005 (pulsar planets not included). The diagram at right is a detail of the diagram at left, between 0 and 2 M_J. The masses of Jupiter, Saturn and Neptune are shown by arrows (mass of Saturn = 0.3 M_J; mass of Neptune = 0.05 M_J).

which soon blow away the disks of gas and dust surrounding them at birth. Without a disk, gas and dust no longer remains to provide material for planets to grow to more than a few times larger than Jupiter. So, here is another mystery for us to solve. How do the heavyweight members of the family of exoplanets come to be?

When we are dealing with even greater masses, the word 'planet' does not really apply. There are a few interlopers within the exoplanet family: the failed stars known as 'brown dwarfs'. Ordinary stars like the Sun create their energy by the fusion of hydrogen into helium. Fusion in brown dwarfs involves deuterium – a heavy isotope of hydrogen with a nucleus comprising one neutron and one proton. Theory predicts that this process will be triggered in bodies of mass 13 M_J. If the effect of inclination is taken into account, any exoplanet with a minimum mass greater than 8 M_J is a suspect. About ten candidates present themselves, so 'brown dwarf contamination' is likely to be limited.

Can we extrapolate our observations to include planets of very low mass? We do not know if it is possible, but let us try anyway. We ought to be finding about 100 times as many planets of 0.1–30 Earth masses than in the range 1–2 Jupiter masses. So, a rich harvest awaits us.

3.3 HOT JUPITERS AND PEGASIDS

At least one in three exoplanets orbits closer in than Mercury

The first exoplanet discovered, orbiting the star 51 Pegasi, caused a sensation when it was revealed that its orbital period was incredibly brief: 4 days! However, 51 Pegasi b is an extraordinary planet only when compared with those in our

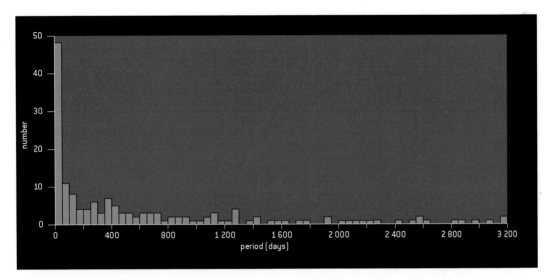

The orbital periods of exoplanets are often short! The periods of the 161 exoplanets known by July 2005 are mostly short: half of them have periods of less than 267 days, and therefore orbit their stars at mean distances less than 0.8 times that of the Earth from the Sun. Theories of the formation of planetary systems suggest that these planets are too massive to have formed *in situ*. Once a large planetesimal has formed, it grows by 'sweeping up' all the matter located in its path. The final mass is *grosso modo* that of all the matter in a ring centred on the star and passing through the planet. However, at 1 AU from the star this ring does not contain sufficient material to form a large, Jupiter-like planet. So, the planet must have formed further out and thrn migrated inwards.

own solar system. Within the family of exoplanets, at least a third have periods shorter than that of Mercury, which orbits in 88 days. Admittedly, these 'Pegasids', as they are known, are still relatively rare; but even so, around twenty known exoplanets have orbits of fewer than 4.2 days. The record-holder is OGLE TR 56 b – a planet of 1.45 M_J, discovered by the transit method, which takes only 1.2 days to revolve around its star. Kepler's third law enables us to calculate the distance between this planet and its star: just 0.02 AU. Astronomers reckon the surface temperature of OGLE-TR-56 b to be as much as 1,600° C – a truly lethal world. 51 Pegasi b may be a 'hot Jupiter', but it seems that there are also 'scorched Jupiters'.

Might we find planets even closer in? Researchers are divided on this question. At a distance of only hundredths of an AU from a star, not only is such a planet hot, but is subject to the irresistible pull of gravity of the star. It is possible that planets so close to their stars have not been found because there are none, as any finding itself in this position will have already been devoured by the star. But the question remains: why do we observe so many planets at around 0.05 AU from their stars? Are they in stable orbits, or merely on their way to being swallowed up by those stars? Astronomers cannot surmise how massive planets can form at distances within a few AU from stars; they must be forming further out, and then

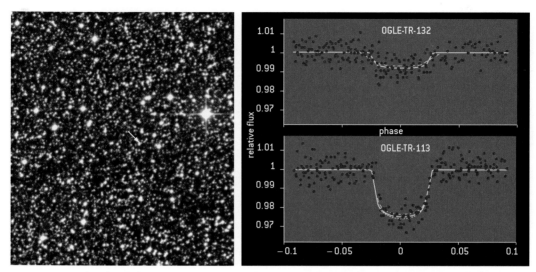

(Left) The field of the star OGLE-TR-113 — a star a little more massive and hotter than the Sun. It is 6,000 light-years away. The OGLE programme, using the transit method, has revealed three planets with orbital periods of less than 2 days. The method measures the size and period of the planet, but not its mass, and complementary radial-velocity measurements are necessary. As the stars studied are very distant, these observations are possible only through major instruments such as the ESO's Very Large Telescope (VLT). (Right). Light-curves from observations of the passage of a planet across the stars OGLE-TR-113 and the star OGLE-TR-132. The radial-velocity curves of these stars were based on observations with the Kueyen telescope of the VLT, which allowed the determination of the masses of the planets detected by the transit method.

migrating inwards. To cause such migrations, various processes may be at work, such as interaction between the planet and the protoplanetary disk in which it formed. The next (difficult) question which arises, though, is this. Once a planet has begun its migration, what is there to prevent it from falling into its star? Are all the close-orbiting planets so far discovered on their way to inevitable destruction? We must be thankful, it seems, that our own solar system is more stable! Computer models suggest that very massive planets are less likely to migrate than their smaller siblings. There are some indications favouring this idea, if we compare the masses of exoplanets to the semimajor axes of their orbits. Hot Jupiters with short periods are not very massive. Also, at distances greater than 1 AU, only planets more massive than Jupiter are found. So, not all exoplanets are scorched. Periods observed range from 1.2 days to 4,500 days, with half the population orbiting their stars in fewer than 300 days. It is difficult to estimate what bias might be introduced by this method of observation; for example, by the timing of the acquisition of data. It is clear that with time the number of known planets in longer orbits will increase. The detection, in 2002, of a planet with an orbital period of 14 years might seem difficult to understand, as before 1995 there were no observing programmes. However, in this case the planet in question is the third to be detected in the vicinity of the star 55 Cancri, and the adjustment of the radial velocity curve was more effective given a three-planet system, with the third in a long orbit.

3.4 SOMEWHAT ECCENTRIC PLANETS

... though hot Jupiters prefer circles

The high frequency of elliptical orbits is a striking feature of the population of exoplanets. We cannot invoke some defect of the observing process to explain this: distinguishing an elliptical orbit from a circular one poses few problems, whether it is the radial velocity or the transit method which is in use. For any

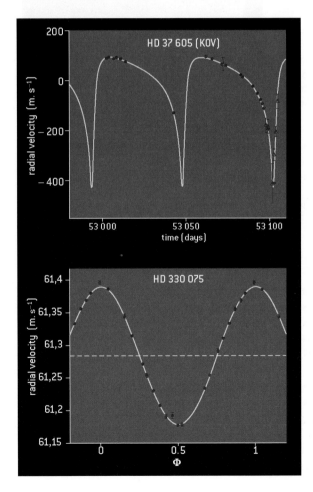

Radial velocities and orbital ellipticities. The radial-velocity method can easily distinguish circular from elliptical orbits. Planets in circular orbits cause regular, symmetrical variations in velocity, expressed by sinusoidal curves. Such is the case with the planet HD 330075 b, discovered by HARPS (lower diagram). High eccentricity produces sharp dips in the velocity curve, as seen in the upper diagram for the star HD 37605. Its companion planet, with orbital eccentricity of 0.74, was discovered in 2004 by the Hobby-Eberly telescope.

Mercury – an eccentric of the solar system. Among the planets of the solar system, only one – Mercury – has an eccentricity greater than 0.2. Mercury is the least well investigated of the terrestrial planets, having been visited by only one spaceprobe, Mariner 10 (in 1974–75), which charted 45% of the planet's surface. A few years from now we will have a much better picture of this planet when NASA's Messenger mission, launched in 2004, enters into orbit around it in 2011. Later, ESA's BepiColombo spacecraft – due for launch in 2013 – will study Mercury. Due to tidal effects caused by the Sun, Mercury completes three rotations on its axis in two Mercurian years (similarly, the Moon completes one rotation during one revolution about the Earth, which explains why it always presents the same face towards us). Such tidal effects are probably significant in the case of all exoplanets orbiting near their stars, and resonances involving rotation and revolution must be common. On a 'Mercury' with a 3:2 resonance and an eccentric orbit, the apparent path of the Sun would seem a little bizarre. As the Sun climbs into the sky it appears to grow bigger, and when it reaches the zenith it stops, turns back on itself, and stops again before descending towards the horizon, 'shrinking' as it goes.

given orbital period, as the orbit becomes more elliptical, the velocity of the planet at periastron (the nearest point to the star) increases, and so detection becomes easier. At the same time, however, the planet spends less time at periastron, and the chance of missing this episode increases. Simulations suggest that the two effects compensate for each other. In summary: of the 185 objects of which the orbital eccentricity is known, at least half have orbits more eccentric than any found in the solar system. Only a few comets have orbits of similar eccentricity.

Why should this be problematic? Again, it seems that the solar system is a special case, and that our theories of its formation do not apply in the case of other planetary systems. In the protoplanetary disk that was the cradle of the solar system, planetesimals grew by accretion from smaller objects, dust and ice (see p. 43). The orbits of these particles were undoubtedly highly elliptical; and the more elliptical their orbits, the more probable that they would intersect the

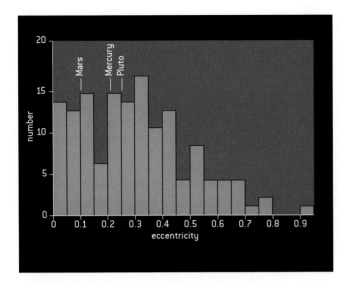

Eccentricities of exoplanets: from circle to flattened ellipse. The orbits of more than half the known exoplanets are more elliptical than that of Pluto. Ellipticity is not the rule, however, and more than twenty exoplanets have practically circular orbits. All the hot Jupiters fall into this category; and are the closest to their stars. The arrows indicate the eccentricities of Mercury (0.206), Pluto (0.246) and Mars (0.093).

orbits of other particles. This natural selection rapidly eliminated the planetesimals in elliptical orbits, eventually leaving only bodies in near-circular orbits, arranged tidily around the Sun. However, this is not at all what is observed when we look at stars other than the Sun.

There are, of course, mechanisms that can be introduced to explain the elliptical orbits: for example, interactions between numbers of planets. But we then have to understand why such mechanisms have not come into play within our planetary system. The eccentricity of an orbit is an important factor in deciding the habitability of a planet. On Earth, long-term climatic variations and glaciations are linked not only with the changing inclination of our planet's axis, but also with variations in the shape of its orbit, its eccentricity and Earth's distance from the Sun at any given time in the cycle of the seasons. Eccentric orbits are common throughout the family of exoplanets, whatever their size, and whether their orbits are close to or far from their stars. Only in the case of planets very close in are all orbits near-circular, and all objects with periods less than 6 days have eccentricities below 0.1. This is easily understood. At such short distances, stars produce enormous tidal effects on these planets. The resulting deformations are considerable, but are stable if the planet's orbit is circular. Planets with initially elliptical orbits will be distorted as the tidal effects dictate, and the dissipation of energy involved will cause the planet to be locked into a synchronous rotation as it revolves in a more circular orbit, as is the case with the Earth's Moon.

3.5 PLANETARY SYSTEMS

Exoplanets *en famille*

Twenty-one of the 200 planetary systems so far discovered are multiple systems in which two or three planets have been detected, and even four in the case of 55 Cancri. This is, perhaps, not many when compared with the number of members of the Sun's planetary family, but we must be mindful of the considerable limitations of current detection methods. If we were situated a few light-years from the solar system and had begun to observe it fifteen years ago with instruments such as ELODIE, we would just about have been able to detect Jupiter. It is very probable that systems of planets are the rule rather than the exception. For once, we can say that our solar system is not a special case, for it seems that planets occur more often in families than alone.

In a few known planetary systems, the orbits of the planets are in a particular configuration known as resonance. The ratio between the orbital periods of their two planets is equal to the ratio of two small whole numbers; for example, 3:2. In the cases of HD 82943 and GJ 876, the outer planet orbits the star in a period almost twice as long as that of the inner planet. With 55 Cancri, the ratio is almost 3:1. Such resonances are well known within the solar system. Pluto and Neptune orbit in a 3:2 resonance. The phenomenon is almost universal in the saturnian system with its many rings and satellites: for example, the satellites

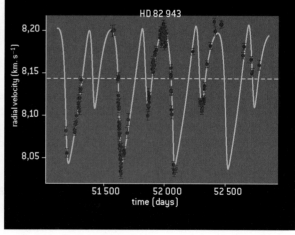

Saturn's satellite Titan, imaged by the Cassini spacecraft. Titan is in a 4:3 resonance with another of Saturn's satellites, Hyperion, which is a rare example of a chaotically rotating body in the solar system.

In the system of HD 82943 the orbital period of the outer planet (c) is 444 days – double that of the inner planet (b). The two planets are in a 2:1 resonance. The radial-velocity curves assume the shape characteristic of the presence of two planets. (After data from the Geneva Observatory.)

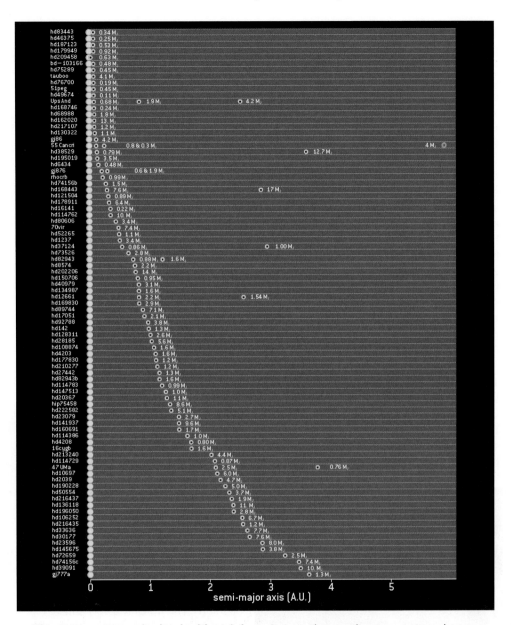

Planetary systems: single, double, triple... Among the exoplanetary systems known there are 21 which have more than one planet; and to these must be added the system of pulsar PSR 1257+12, with its three planets (and possibly a fourth, unconfirmed). Planets in these multiple systems greatly resemble 'solitary' planets in characteristics such as mass, period and eccentricity; but this is hardly surprising, as the latter are doubtless 'solitary' only because current instruments are not sufficiently powerful to detect their neighbours. On the graph, the figures give the masses of the first exoplanets to be discovered as a function of distances from their stars. (After www. exoplanets.org.)

Titan and Hyperion are in a 4:3 resonance. To this we can add the multiple planetary system of the pulsar PSR B1257+12, with at least three planets; of which two, planet b (4.3 Earth masses) and planet c (3.9 Earth masses), are in orbits with a resonance of almost 3:2. It is also interesting that these planets have been shown to be orbiting in practically the same plane. Even if the neutron star associated with the phenomenon of a pulsar is nothing like the Sun, this suggests that the process of planetary formation is the same with both types of object, with an initial disk of gas and dust rotating around the star.

The observed resonances show that dynamical effects may be important in these multiple systems. If two planets have achieved resonant orbits, the gravitational interactions between them repeat at the same stages of their orbits and may be amplified. It is not impossible, therefore, that one of the planets might eject the other from its orbit. Those who specialise in celestial mechanics are very interested in this question of the stability of multiple systems, and are delighted to have something outside the solar system to study! The results from the planets in a 2:1 resonance are, however, slightly disturbing. Over a million years it is quite possible that the larger of the two might find itself ejected by the smaller.

3.6 CANNIBAL STARS?

The probability of detecting a planet is greater if its star contains more of the heavy elements

Patience is required in the search for exoplanets. Thousands of stars must be studied in order to detect just a few planets. Astronomers have patience to spare, but they soon realised that certain stars provide happier hunting grounds than

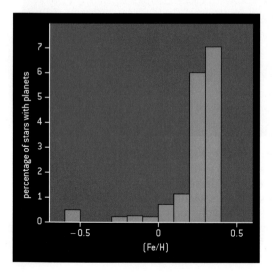

Iron to make planets. The probability of detecting planets around a solar-type star (vertical axis) is very dependent on the amount of heavy elements in the star's atmosphere: here, the quantity of iron compared to that of hydrogen, with reference to the solar value (logarithmic units). The Sun lies at [Fe/H = 0]. Only a very small number of stars with metallicities less than or equal to the Sun's are known to have planets. As metallicity increases, so does the planet count – and steeply. The abrupt decrease in the detection rate below [Fe/H = 0.4] is simply due to there being no iron-rich star in the vicinity of the Sun.

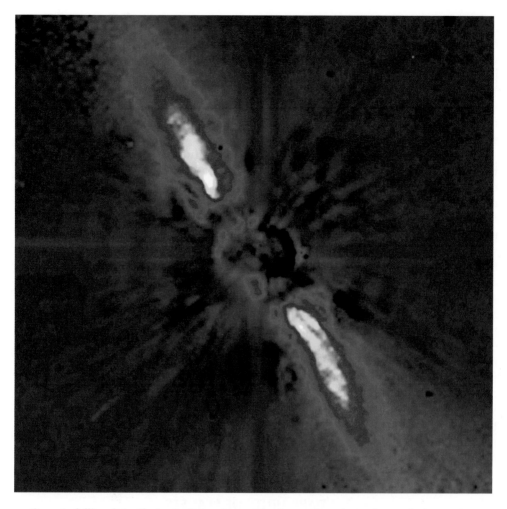

Comets falling into their star? Spectrometric observations have shown that comets constantly fall into their central stars. Here is β Pictoris, in the southern sky, as seen by the 3.6-metre telescope of the ESO. The star is surrounded by a disk of gas and dust. This disk is certainly evidence of the intense activity associated with young planetary systems. The image of the central star has been masked using the coronagraphic technique, allowing us to see the disk at less than 25 AU from the star.

others. The abundance in stars of chemical elements heavier than helium (which astronomers collectively and idiosyncratically call 'metals') seems to be a key parameter. The detection rate climbs by a factor of 7 in studies of stars with twice as many heavy elements as there are in the Sun. This is not a result of some observational bias, as stars of metallicity equivalent to the Sun's are much more common than those with more metals.

Only a handful of stars with lower metallicity than the Sun's are known to have planets. What can this result mean? We know that terrestrial planets are made of heavy elements: silicon, oxygen, nickel and iron. The giant planets which we observe orbiting other stars are undoubtedly gaseous, but their hydrogen–helium atmospheres probably conceal cores rich in heavy elements. These are therefore an essential ingredient in the formation of planets as we know them. It seems likely, then, that planets form more readily in environments rich in heavy elements – elements also found in the atmospheres of their stars, which were born of the same material.

However, we also know that planets can (and, indeed, must) migrate towards the interior of planetary systems. The more readily the process of migration begins – perhaps by interaction between planets and the protoplanetary disk – the less probable it is that the migration can be halted. Could it be that some of those planets situated at just hundredths of an AU from their stars end up by plunging into them? If this is so, then the rocky material of those planets will be vaporised in the star's atmosphere. But this cannibalism of planets does not seem to be the dominant process. On average, there is 1.8 times as much iron in a star accompanied by planets than in a star with none. If we calculate how much rocky material is required to explain this difference, we realise that stars would have had to absorb tens of planets. Now, observations confirm that protoplanetary disks have a lifetime of less than ten million years. The swallowing up of just a few planets by a star might occur in such a short period, but it is much less likely that tens of planets could suffer the same fate during that time. We have learned an important fact: that planets form more readily in an environment rich in metals. Models simulating the formation of planetary systems will have to take this into account. It will also be necessary to examine the possibility of the formation of planets in environments of medium metallicity – the best example being our solar system, as the Sun resides in the range of metallicity where the probability of finding planets is quite low!

3.7 DARK HINTS OF PLANETS

... or, how to measure the radius of an exoplanet

At least ten research teams in various countries have taken upon themselves the task of finding exoplanets by means of the transit method. Analysis of the light-curves of the millions of stars studied has revealed hundreds of candidate events which might be ascribed to exoplanets. Velocimetry is then introduced, using powerful instruments such as the Very Large Telescope (VLT) of the European Southern Observatory. With this method it can be determined whether these events are caused merely by binary stars or by transits of exoplanets. Big telescopes are essential, because most of the candidate stars are very distant and faint. Their identity as star–planet systems can be confirmed, and the masses of the planets estimated. This tireless hunt has begun to bear fruit, and ten of the

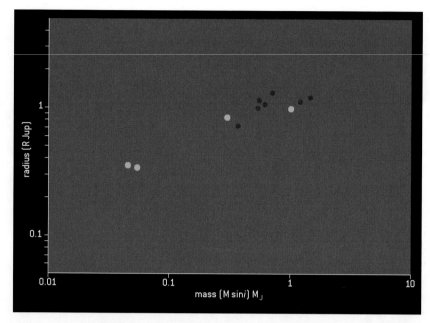

Mass/radius relationship for exoplanets detected by the transit method (in red), and the giants of the solar system (in blue). The exoplanets resemble solar system giants.

many suspects are now known to have planets. Painstaking dedication to this work has proved itself worthwhile, and reveals two important factors. The first, as we have seen, is the radius of the planet, deduced from the relative dimming during the transit being proportional to the square of the ratio of the radii of the planet and the star, $(R_p/R_*)^2$. The second is that the planetary system must be observed practically edge-on, else no transit will be seen. Remember that the radial velocity method determines only the minimum mass of the planet, $M \sin i$. For all the planets detected by the transit method and subsequently confirmed by the radial velocity method, we know that i is close to $90°$ and $\sin i = 1$. We therefore have two quantities: the mass and the radius of the planet. With these, we can embark upon some physics!

We can say for certain that these exoplanets are giants, as their radii are comparable to that of Jupiter; and, like Jupiter and Saturn, they are composed mainly of gases. Their mean density is of the order of 0.8 gm/cm^3 – less than that of water, and far less than that of the terrestrial planets (the mean density of Earth is 5.5 gm/cm^3). If we could find a sufficiently large bowl of water and put the exoplanets into it, they would all float, with the exception of OGLE-TR-113 b! The giants of the solar system are a little denser, but only Saturn is, overall, less dense than water. In the final analysis, all this strongly suggests that exoplanets are similar in constitution to the giant planets of the solar system. There do not seem to be any rocky 'super-planets'.

Giants, and very hot giants at that... The nine planets in question share the

ESO's GIRAFFE spectrograph obtains high-resolution spectra of several objects, simultaneously. The transit method requires complementary velocimetric observations with powerful telescopes in order to establish whether candidate stars really have exoplanets, as many of these stars are very distant and faint.

characteristic that they are very close to their central stars, with mean distances ranging from 0.022 AU to 0.047 AU, and orbital periods between 1.2 and 4 days. The planet that is closest to its star must have a surface temperature of about 1,600° C! Because all these planets are 'hot Jupiters' there is certainly an effect of selective bias in the method of detection, and we have seen that the probabilty of a transit varies as R_*/a, where a is the radius of the orbit. These planets are at just a few tens of stellar radii from their stars.

3.8 CLOSE-UP OF A HOT JUPITER: HD 209458 b

Is this planet evaporating?

HD 209458 b is also known as Osiris: two names for an exoplanet which has become, for astronomers, *the* exoplanet, most of whose physical and orbital parameters have been determined. This exoplanet, 47 parsecs (153 light-years) away in the constellation of Pegasus, is obliging enough to have an orbit which presents itself edge-on across its star, as seen from Earth. It is therefore susceptible to observation by both transit and radial velocity techniques.

Since it was first observed in 1999, this 'star among stars' and its companion have been scrutinised as no other, from Earth and by means of the Hubble Space Telescope. We therefore know quite a lot about planet HD 209458 b. Its orbital period is 3.52474 days, and it is 0.045 AU from its star. We know that its radius is 1.347 R_J $\pm$ 0.06 R_J, and its mass is 0.69 M_J $\pm$ 0.05 M_J. These are indeed remarkable findings for a body so far away.

Although HD 209458 b is less massive than Jupiter, it has a greater diameter. This set astronomers thinking. Might this planet, so close to its star, be in the process of losing its atmosphere? Could they learn more about that atmosphere? Visible and ultraviolet spectroscopic analyses of the star were carried out, both during the transit and when no transit was occurring. In this way, various teams sought possible spectral signatures of the planet itself – and they were successful. Evidence of the presence of sodium atoms suggested that clouds of sodium were present deep within the planet's atmosphere. Moreover, absorption features due to hydrogen, carbon and oxygen were detected. These were more intense than might have been expected, leading to the conclusion that the atmosphere is

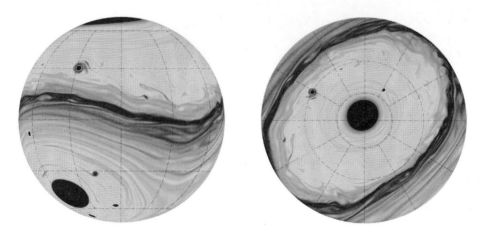

Computer simulations of atmospheric circulation on exoplanet HD 209458 b. (After Cho, Menou *et al.*, *Ap. J.*, **587**, L117.)

An artist's impression of the oxygen/carbon atmosphere surrounding exoplanet HD 209458 b. Alfred Vidal-Madjar and his team at the Paris Astrophysical Institute were the first to detect the signal of oxygen and carbon in the atmosphere of an extrasolar planet. The nearness of the planet to its star is thought to cause the evaporation of the atmosphere into space, with the radiation pressure of the star forming an immense 'tail' reminiscent of the train of a comet.

indeed expanding, or even evaporating! The proximity of the star, just 7 million km away (0.05 the distance from the Earth to the Sun), is causing the evaporation. The radiation pressure of the star pushes the evaporated gas away from the planet, creating an immense 'tail' like that of a comet. Computer models suggest that in certain circumstances the whole atmosphere could be stripped away over a few billion years, leaving behind a small rocky remnant.

Thus the name Osiris. The dismembered body of this Egyptian deity was scattered all over Egypt in an attempt to deny him immortality.

Some astronomers have undertaken studies of the 'climate' of HD 209458 b; a singular kind of climate, to be sure – not only because of the nearness of the star, but also because it is certain that the planet always keeps the same face turned towards it. Intense tidal effects have locked the star–planet system into a configuration like that of the Earth and its Moon; and this is probably the case with all exoplanets with orbital periods of a few days or less. The difference in temperature between Osiris's 'day' and 'night' sides is likely to engender violent winds, of the order of several thousand km/h. In spite of these winds, which would tend to homogenise atmospheric temperatures a little, the 'daytime' side no doubt experiences a scorching 1,500° C, some 600° hotter than the night side. These results are obviously much debated, as meteorology is a complex science even when applied to the Earth!

3.9 HOW MANY STARS HAVE PLANETS?

Planetary systems are common – but how common?

The detection of planets utilising the radial velocity method cannot be referred to as 'highly productive'. As we have seen, if the Sun and its planets were to be among those so far studied by this method, it might still be some years before we could detect Jupiter. Saturn might be found, but not for about twenty years. Nevertheless, the list of known exoplanets has now passed the 200 mark.

So, is it likely that many stars have planets? Initially, the answer is 'no'. Only a small percentage of stars observed with the velocimetric method have planets, and the transit method has produced far fewer results. In reality, however, the proportion of stars possessing planets is certainly much greater. There are several explanations for this. First among them is that we are not detecting all types of

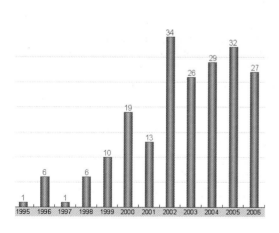

Annual numbers of discoveries of exoplanets since 1995 (including the first seven months of 2006 only). In spite of the introduction of new instruments in the last few years (up to and including 2006), the radial-velocity method is unable to reveal hundreds of extra planets in the future. Nearly all observable stars are now studied, and the short-period planets are already detected. Long-period planets are found as studies proceed, but fewer of these have been detected. Fewer planets with periods between 500 and 510 days have been discovered than planets with periods between 0 and 10 days.

Our galaxy, the Milky Way – seen here above one of the telescopes of the Australian National University – contains approximately 300 billion stars. If we suppose that only one in every 100 of these stars has a planetary system, then there must be several billion such systems in our galaxy!

planet. There are the small ones, whether or not they orbit close to their stars; and large ones, if they are in remote orbits like Jupiter's. It may be no coincidence that of those stars which have been observed for the longest period of time (some fifteen years), more than 10% are known to possess planets.

There is no kind of star, even of solar type, which is easy to observe. A small degree of variability, involving just the odd eruption, will contaminate the spectral signature of a star and considerably increase the difficulty of detecting any planets. Moreover, if we restrict ourselves only to stars of very limited variability, one in four will yield results. Other targets offering a high probability of discovering exoplanets are those stars with a higher than average content of heavy elements. At a metallicity level twice that of the Sun, one star in four has a detectable planetary companion. In summary: of a sample of ordinary solar-type stars, only 5% will possess planets; and in carefully selected subsamples the figure can rise to one in four. However, the method used is not sensitive. Finally, there is nothing to prevent us from imagining that nearly all these stars have planets, and certain researchers already consider this to be so. Yet how are we to discover all those planets? In time, the radial velocity method will reveal all 'Jupiters', 'Saturns' and 'Neptunes' in long orbits, but Earth-like planets will be forever outside their range. These questions have given rise to projects, both on the ground and in space, with the avowed object of finding exoEarths.

It is worth noting that, effectively, nearly all the stars which can be observed by this method are being observed, and often by more than one team. Here we have some thousands of stars under scrutiny, of spectral type similar to that of the Sun and of suitable magnitudes. The sample can be enlarged if fainter objects are included, through the use of telescopes more than 2 metres in aperture. This is the aim of the HARPS (High Accuracy Radial Velocity Planet Searcher) spectrograph, currently mounted on a 3.6-metre telescope.

3.10 FAILED STARS OR SUPERMASSIVE PLANETS?

Planet or brown dwarf: a question of mass

Star or planet? This is not as easy a distinction as one might think, although theoretically a star produces energy from thermonuclear fusion reactions and a planet does not do so. The Sun shines because, like most other stars, it transforms hydrogen into helium, while the light of the planets accompanying it is merely reflected sunlight. Let us start from the beginning, from the birth of a star. A small 'knot' of matter within an interstellar cloud is drawn inwards upon itself by its own gravity. Its core becomes more and more dense, and ever warmer. If the initial knot of matter is sufficiently large, the core of the resulting protostar will become hot enough to switch on the thermonuclear reactions. The fusion process which turns hydrogen into helium requires a temperature of about 10 million degrees, implying a mass of the order of 0.08 times that of the Sun (about 80 times the mass of Jupiter). If this value is attained, a star will be born. The smallest hydrogen-burning stars are the red dwarfs.

Below this mass the hydrogen fusion reaction will not proceed, but deuterium – a heavy isotope of hydrogen – may instead come into play. We

The Trapezium, in the Orion Nebula. The points of light seen here, with the exception of the main (young) stars, are brown dwarfs and 'drifting' planets unattached to any star. This false-colour photomosaic is based on a combination of infrared and visible-light images obtained by the Hubble Space Telescope.

are now in the realm of brown dwarfs, about which it might be said: 'When is a star not a star?'

Brown dwarfs shine feebly (0.0001 the luminosity of the Sun), but have lifetimes of billions of years. If a body has less than thirteen times the mass of Jupiter (about 0.01 of the mass of the Sun), then its core temperature will not be high enough to trigger the fusion of deuterium. Any object below this crucial mass of 13 M_J will therefore remain a planet.

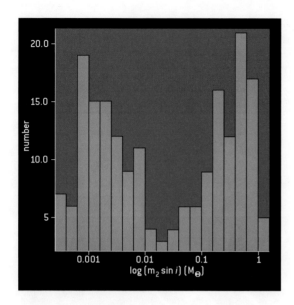

The number of companion objects of a solar-type star as a function of mass ($M \sin i$), from planets (<0.01 solar mass) to ordinary stars (>0.08 solar masses). There are very few objects in the 'brown dwarf desert' between 0.01 and 0.08 solar masses. (After Santos *et al.*, *A. & A.*, **392**, 215 (2002).

This definition, however, might be considered a little bizarre. Surely a planet is, by its very nature, in orbit around a star. Could there be planets drifting in interstellar space, not associated with any other body? Some astronomers claim to have discovered 'floating planets'; but in the absence of any accurate measurements of these bodies it is difficult to reach a conclusion.

Might a planet merely be a failed brown dwarf – a kind of 'doubly failed' star? Observations suggest that brown dwarfs and planets are rather different, and evidence of this is the 'brown dwarf desert'. When we investigate the types of companion that solar-type stars have, we find far fewer objects in the range 0.01–0.08 times the mass of the Sun, between the zone of planets (below 0.01 solar mass) and the zone of stars (more than 0.08 solar mass). This clear separation of stars from planets reminds us that the two types of body are formed in different ways – stars via gravitational collapse, and planets by accretion of matter onto a rocky embryo within a disk of gas and dust.

But beware! Some of the most massive exoplanets may indeed be brown dwarfs, as the radial velocity method determines the mass only to within a factor of the inclination.

4
What do we learn from our own solar system?

We think that we now understand how our solar system formed, with its two types of planets in their concentric, near-circular orbits, all travelling in much the same plane. But this model of formation is not echoed by what we have discovered about new worlds...

An artist's impression of the solar system, based on data from NASA space missions.

4.1 EARLY THEORIES: THE SEVENTEENTH CENTURY ONWARDS

Like neighbouring protoplanetary systems, the solar system formed within a disk

What is the Earth's place in the Universe? This question is intimately connected with that of the existence of other inhabited worlds, and is as old as the human race. The Greeks, following the teachings of Aristotle, placed the Earth at the centre of everything. For more than 1,000 years this was the prevalent theory, in spite of the work of early thinkers such as Aristarchus of Samos (who lived around 300 BC) and, many centuries later, Nicholas of Cusa (1401–1463), both of whom proposed a heliocentric (Sun-centred) model. It is to Nicolaus Copernicus (1473–1543) that we owe the accession of the heliocentric system, now known as the Copernican system. In 1543 (the year that Copernicus died) his radical book *De Revolutionibus Orbium Coelestium* was published.

Copernicus's theory, which sat ill alongside the teachings of the all-powerful Catholic church of the time, finally found acceptance thanks to the work of mathematicians and astronomers such as Tycho Brahe (1546–1601), Johannes Kepler (1571–1630), Galileo Galilei (1564–1642) and Isaac Newton (1642–1727), whose calculations and observations scientifically justified the Copernican model. Tycho's observations were profitably taken up by his pupil Kepler, who

The German philosopher Immanuel Kant (1724–1804) and the French scientist Pierre-Simon de Laplace (1749–1827), who formulated the first theories concerning the origin of the solar system.

The Orion Nebula. This nebula is a complex of various nebulae – the brightest among them being M42 (the pink cloud at centre right). M43 is the smaller, round nebula (below centre), separated from M42 by a narrow blue band of gas. M42 and M43 appear pink due to the presence of hydrogen ionised by the young stars within them. To the left is the nebula NGC 1977, which, being non-ionised, reflects the light of nearby blue stars. The Orion Nebula is about 1,500 light-years from Earth.

formulated empirically the famous 'Kepler's laws', describing the motions of the planets around the Sun. Galileo – an early promoter of the telescope for astronomical purposes – confirmed the Copernican theory in new ways, and Isaac Newton, in *Philosophiae Naturalis Principia Mathematica*, propounded the law of universal gravitation – mathematical proof of Copernicus's model.

With the Sun in its rightful place at the centre of its system, the question remained as to how it was made. The first hypotheses concerning the formation of the solar system were proposed by René Descartes (1596–1650). In his *Théorie des Vortex* Descartes imagined the solar system to have been formed from a 'whirlpool' of matter, but this model did not explain why the planets move in approximately the same plane (the ecliptic). Neither did it suggest why their orbits are almost circular, and why they all orbit in the same direction.

Later, Immanuel Kant (1724–1804) and Pierre-Simon de Laplace (1749–1827) introduced the nebular theory to explain these two fundamental characteristics of the solar system. The main aspects of this theory – which were intuitive at the time – are nowadays largely accepted. An initial, contracting gas cloud begins to turn with increasing speed, and finally collapses into a disk. Within this disk form planets, together with their satellites and ring systems (see p. 43).

4.2 HOW OLD IS THE SOLAR SYSTEM?

Clues from measuring radioactivity in lunar rocks and meteorites

What means have we of dating solar system objects? The method used by astronomers was developed during the second half of the twentieth century. It is based on measuring the age of certain radioactive elements which disintegrate very slowly over a period comparable to the lifetime of the Sun. These 'parent' elements are often isotopes* of a stable element, decaying into a different 'daughter' element of the same atomic mass, as a function of time, according to an exponential law. Through measurements of the amounts of the stable initial element and the parent and daughter elements, we can estimate the epoch at which the decay commenced, and thereby the age of the sample in question (for

A fragment of a meteorite – possibly part of the asteroid Vesta – which fell in western Australia in 1960.

A lunar crater. Like those on Mars and Mercury they are evidence of the period of intense meteoritic bombardment which occurred during the first billion years of the history of the solar system.

---

*   Most chemical elements have one or more isotopes; that is, elements with the same number of protons and electrons but a different number of neutrons, giving them a different atomic mass.

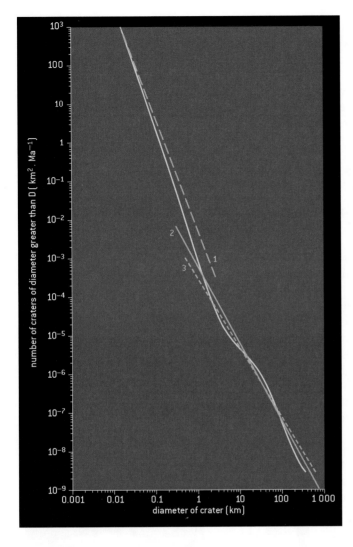

Dating a planet according to the degree of cratering on its surface. The horizontal axis shows crater diameters D, and the vertical axis the number of craters with diameters greater than D. (After Hartmann and Neukum, *SSRV*, **96**, 165 (2001).)

example, an Earth rock, a lunar sample or a meteorite). This method is known as isotopic dating. Often, the elements involved are the pairs argon-40/potassium-40 (^{40}Ar/^{40}K), uranium-238/lead-238 (^{238}U/^{238}Pb) or rubidium-87/strontium-87 (^{87}Rb/^{87}Sr). In the case of this last pair, the amounts of the elements ^{87}Rb and ^{87}Sr are compared with that of (stable) strontium-86 (^{86}Sr), providing an indication of the initial abundance of strontium in the sample tested.

These analyses can lead to some spectacular conclusions. They have shown us that the age of the solar system, as measured in the most ancient parent bodies of meteorites, is 4.56 billion years, $\pm 1\%$. It is thought that the various bodies comprising the solar system – planets, asteroids, comets and so on – formed over a relatively short period of some 100 million years at most. As well as chemical analysis of extraterrestrial samples there is another way of dating planetary surfaces. This involves examining craters upon planets and satellites without dense atmospheres, such as Mercury, the Moon, Mars, the asteroids and the Galilean satellites of Jupiter. The more heavily cratered a surface, the older it must be. By comparing the ages of lunar rocks with the number of craters we can examine the relationship between the age of the surface and the degree of cratering, which can then be applied to other planetary surfaces. It becomes apparent that during the first few hundred million years of the lifetime of the solar system there was an extremely heavy bombardment, which reached its zenith about 3.8 billion years ago. It will be seen that this bombardment was due to a particularly active episode in the life of the young Sun, when it drove the debris of its protoplanetary disk outwards. So, by dating surfaces and analysing extraterrestrial samples we can discover precious information about our very origins.

4.3 LOOKING AT NEARBY STARS

The Sun is halfway though its life: in mass and luminosity, a very ordinary star

Since the Sun is a very unremarkable star in our galaxy, a look at its neighbours may help us to understand its history. Let us concentrate upon very young stars and star-forming regions. These regions are dense, rapidly rotating molecular clouds which will in time collapse, giving birth to a new star. During recent decades the study of the early phases of a star's life has proceeded in leaps and bounds, due mainly to infrared and millimetric astronomy, which allows us to penetrate cold environments opaque to visible light. It has also been possible – especially with the Hubble Space Telescope – to secure images of protoplanetary disks around infant stars.

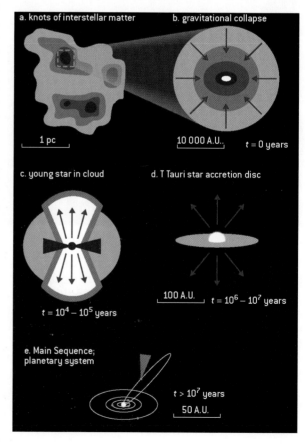

The phases of star formation. A rotating cloud of material collapses into a disk; and at its centre, material is concentrated into a protostar which becomes a star when nucleosynthesis is triggered at a temperature of a few million degrees. This protostar is surrounded by a disk within which grains agglomerate to form planetesimals.

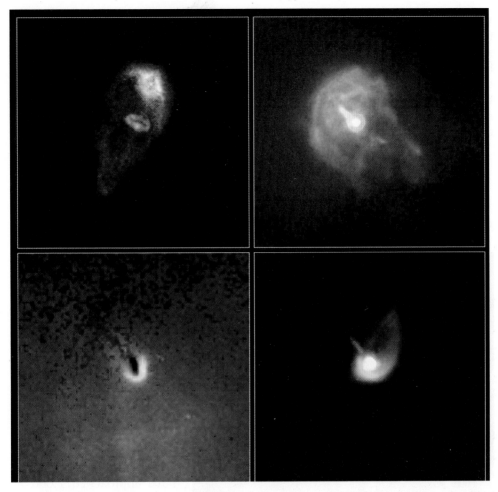

Protoplanetary disks in the Orion Nebula, imaged by the Hubble Space Telescope's wide-field planetary camera (WFPC2).

Certain very young stars are particularly intriguing to astronomers. These stars too are surrounded by disks, and are characterised by intense activity involving the outward impulsion of very large quantities of matter. This active period is known as the 'T Tauri phase' (from the name of the first star around which such activity was observed), which occurs quite soon after the birth of a star, during the first ten million years of its lifetime.

It is reasonable to suppose that the Sun underwent a largely similar process to that observed in the case of nearby stars of about the same mass. The most telling evidence is the existence of the ecliptic plane, a vestige of the original protoplanetary disk, and the fact that the planets orbit in near-circles, all proceeding in the same direction. So, early on a rotating cloud of matter collapsed into a disk; and at its heart, matter was concentrated into a protostar.

When the object's internal temperature reached a few million degrees, nucleosynthesis was triggered, and the object became a star. Within the disk, grains of matter collided to form planetesimals, which later, by accretion, became planets. When the Sun's T Tauri phase intervened, the gas and dust of the protoplanetary disk were expelled. All that remained were protoplanets, or the largest embryonic planets. Later came the era of bombardment, lasting for several hundred million years – ample evidence of which is presented by the crater-strewn surfaces of Mercury and the Moon.

4.4 THE PROTOPLANETARY DISK: THE CURRENT PICTURE

Hydrogen: dominant element of the protoplanetary disk

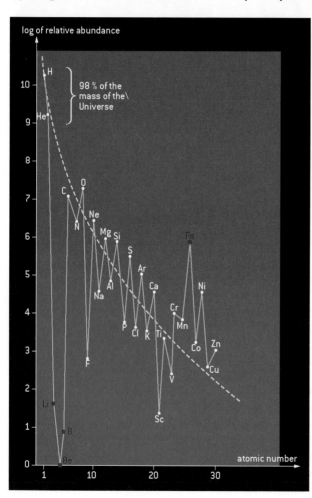

Let us try to picture the protosolar disk shortly after its collapse. Given the positions of the resulting planets it must have extended for several tens of AU. Its total mass was a small fraction (perhaps a tenth to a hundredth) of that of the protostar, and its density and temperature decreased from the centre outwards. Near the Sun, temperatures might have approached 2,000 K

A curve showing the abundance of chemical elements in the Universe. Only elements lighter than zinc are included. These values are measured principally in the Sun (with the exception of deuterium) in the case of relatively light elements – C and O – and in meteorites in the case of the heavier elements. Li, B and Be are very rare, while Fe (in red) is very abundant. (After C. Allègre, 1985 (modified).)

An artist's impression of a brown dwarf surrounded by a protoplanetary disk.
NASA's Spitzer Space Telescope has found a disk like this around the very-low-mass
brown dwarf star OTS 44, which is about fifteen times larger than Jupiter.

(we shall later see how this is known), while at distances between 30 and 50 AU they would have been only 100 K at most. The whole disk would slowly cool down as time passed.

What did this protoplanetary disk contain? All the elements present in the cosmos were there, in the relative abundances with which we are familiar. Hydrogen and then helium were the most abundant. Both of these were in gaseous form, since their condensation temperatures are so low. Then, in decreasing order of abundance, there were oxygen, carbon and nitrogen – the first elements to be formed after helium during stellar nucleosynthesis. We shall see how these can be found in solid or gaseous form according to the ambient temperature within the disk; that is, according to their distance from the Sun at a given time. Then come the heavier elements, also synthesised in stars, but in smaller quantities: sodium, magnesium, aluminium, silicon, phosphorus, sulphur, chlorine, calcium and so on. Iron, with its very stable configuration, is particularly abundant, given its high atomic mass. All these 'refractory' elements were present in solid form within the protosolar disk. They can be found in meteorites and lunar samples.

Note that other heavy elements, likewise made inside stars, are also present in the disk in gaseous form. These are the rare or noble gases which, together with helium, are neon, argon, krypton and xenon. Because they are chemically inert they are precious indicators of the evolution of the atmospheres containing them.

4.5 THE ICE LINE

The formation of water ice marks the boundary between terrestrial planets and giants

It was solid particles which, within the protosolar disk, came together in multiple collisions to form embryonic planets or planetesimals. Later, these gave rise to the planets and their attendants, to asteroids and to comets. To understand the mechanism of their formation we must first identify the solid constituents of the protosolar disk.

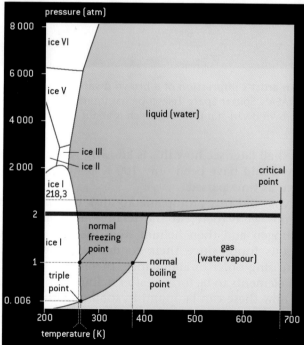

The phase diagram of water. The curves of the diagram represent boundaries between the different states of water, and function of temperature and pressure. Roman numerals refer to the different types of crystalline ice. Note the change of scale on the vertical axis at 2 atmospheres.

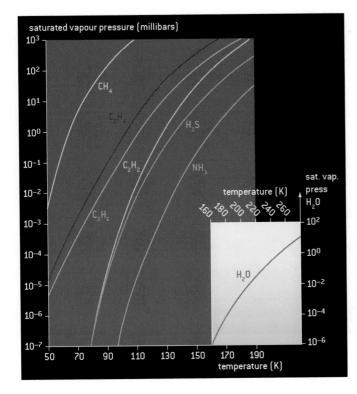

Phase diagrams (saturated vapour pressure/temperature) of some condensable gases. The inset shows the saturated vapour pressure of water, as a function of temperature. Water vapour is the gas which, at a given pressure, condenses at the highest temperature. (After S.K. Atreya, *Atmospheres and Ionospheres of the Outer Planets and Their Satellites*, Springer-Verlag, 1986.)

As we have seen, elements of atomic mass above approximately 20 – with the exception of the rare gases – are refractory. What of the lighter and therefore more abundant elements carbon, nitrogen and oxygen? They enter into association with hydrogen to form methane (CH_4), ammonia (NH_3) and water (H_2O). They can also form CO, CO_2, HCN, H_2CO and other compounds, and atoms of hydrogen combine to form the H_2 molecule.

In what form are these molecules found? Near the Sun they are all in gaseous form, and cannot have contributed to the composition of the solid embryonic bodies which became planets. However, at a few AU from the Sun the temperature was low enough for all of them, except molecular hydrogen, to exist as ice. As a function of the distance from the Sun, the first molecule to condense out is H_2O, and, with increasing distance, as the temperature falls, there follow NH_3, HCN, CO_2, H_2CO, CH_4 and so on (see diagram above).

There is therefore a critical distance from the Sun beyond which the matter contained within the protoplanetary disk will occur mainly in solid form: the 'ice line'. At present it is situated between 1 and 2 AU from the Sun, although during the early history of the disk, before cooling set in, it lay further out.

How did planets form close to the Sun? Remember that the mass contained within the refractory elements, which are heavy and scarce in the Universe, is small compared with that of the ices containing carbon, nitrogen and oxygen,

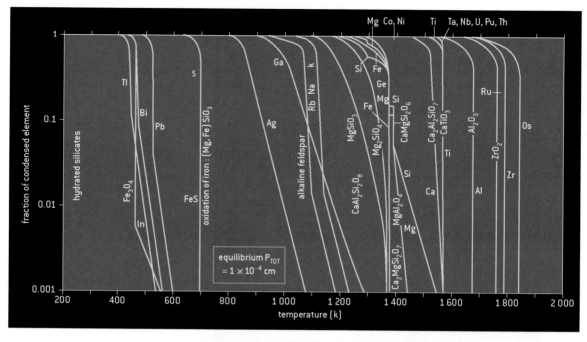

Condensation sequence for refractory elements for a gas of solar-type composition. (After Grossmann and Larimer, *Rev. of Geophysics and Space Physics*, **12**, 71 (1974).)

the most abundant elements after hydrogen. So, the mass of material available for planet formation was limited. Models predict the emergence of only a few protoplanets, of masses lower than or similar to that of Earth, and with high densities (3.9–5.5 g/cm^3). This is precisely what we find in the inner solar system.

Beyond the ice line the amount of available solid matter within the disk was sufficient for the formation of large icy nuclei, perhaps a dozen times more massive than the Earth. Theory shows that their gravitational fields would have been strong enough to draw in surrounding protosolar gases, essentially hydrogen and helium. New planets were born. They were very massive and extremely large, with low densities (0.7–1.7 g/cm^3) – today's giant planets.

4.6 TERRESTRIAL AND GIANT PLANETS

Can the differentiation between terrestrial and giant planets be found in extrasolar systems?

The model of planetary formation by accretion around solid particles leads naturally to the existence of two distinct classes of planet: terrestrials and giants. What is the composition of their atmospheres? In order to investigate this

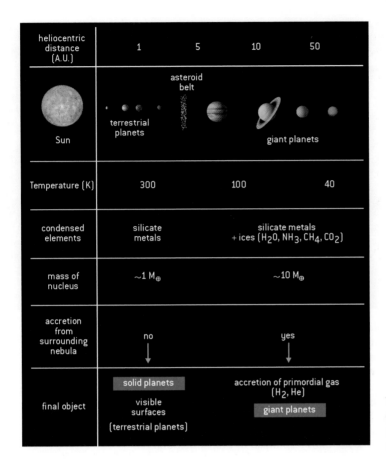

heliocentric distance (A.U.)	1	5	10	50

The two formation processes of the solar planets (terrestrial and giant).

question, let us return to the chemical composition of the protoplanetary disk. Thermochemical models predict that within giant planets – where the pressure is relatively high but temperatures are low – carbon and nitrogen occur preferentially in the form of methane and ammonia. Indeed, this is what we observe in the case of the giants. Within the inner solar system, thermochemical equilibria involving CH_4, H_2O and NH_3 evolve to form CO, CO_2, H_2O, N_2 and H_2. Molecular hydrogen is too light to remain trapped in the gravitational fields of the terrestrial planets, and escapes. This is the initial global composition of the atmospheres of the inner planets, and all of them have undergone considerable evolution. We shall return to them later.

The giant planets fall into two distinct categories. Jupiter and Saturn – with masses respectively 318 and 95 times the mass of the Earth – are essentially composed of protosolar gases, and are rightly called 'gas giants'. However, Uranus and Neptune (15 and 17 Earth masses) consist mostly of their initial ice cores, and are 'ice giants'. The chemical composition of the giant planets offers clear confirmation of the model of planetary formation by accretion around a central nucleus.

Sun
Jupiter
Saturn
Uranus
Neptune
Venus   Earth
Mercury   Mars
(an e

terrestrial planets   gas giants   ice giants

giant or Jovian planets

The planets of the solar system fall into two main categories: terrestrials (within 2 AU from the Sun) and giants (beyond 5 AU). The terrestrials (Mercury, Venus, Earth and Mars) are rocky, small, very dense, possess few satellites and, with the exception of Mercury, are enveloped in stable atmospheres. The giant planets are massive but not dense, and have rings and many satellites. The two largest – Jupiter and Saturn – are the nearest of the giants to the Sun and are composed mainly of gas, while the other two – Uranus and Neptune – consist mainly of ice. The last object, Pluto, beyond the orbit of Neptune, and previously considered as a planet, is now seen as one of the largest members of a recently discovered class of bodies: the trans-Neptunian or Kuiper Belt objects.

Now, 2% of the protosolar gas consists of heavy elements – a standard value in the Universe – but in initial nuclei they make up 100% of the material. In accordance with theoretical models, let us take an initial nucleus of 12 Earth masses. Suppose that all the heavy elements are fixed in ices, and that the mixture within giant planets is homogeneous, after the collapse of the protosolar gas. For these planets we can now calculate the excess of the heavy elements *vis-à-vis* hydrogen, compared to cosmic values: 3 in the case of Jupiter, 7 for Saturn, and 30–50 for Uranus and Neptune. These excesses have indeed been measured in these planets, thereby confirming unambiguously the nucleation model.

Returning to the terrestrial planets, we now ask: how did their atmospheres originate? As already stated, protosolar gases do not figure here. Hydrogen and helium are too light to be held by the terrestrials' gravitational fields, which doubtless acquired their atmospheres partly through outgassing and partly through meteoritic bombardment, from the gaseous elements contained within asteroids and comets. The Earth's water in particular is certainly, at least in part, of cometary and asteroidal origin, as may be the prebiotic molecules which were the precursors of life on Earth.

4.7 FROM JUPITER TO NEPTUNE: FOUR DIFFERENT WORLDS

A great variety of giant planets may exist within extrasolar planetary systems

Although they are usually listed together as 'giant planets', the gas giants and the ice giants are very different from each other. Jupiter, at a heliocentric distance of 5.2 AU, is the most massive and the largest, as well as being the giant nearest to the Sun. Its diameter is about a tenth that of the Sun, and its mass about a thousandth of a solar mass. Situated just beyond the ice line, Jupiter was best placed to capture the greatest number of planetesimals, and rapidly grew larger.

How long did it take for Jupiter to form? A few million years, and perhaps less. Its growth must have preceded the dissipation of the protosolar gas during the youthful Sun's T Tauri phase (see p. 72). Observations of nearby young stars reveal that this phase seems to last for not more than about 10 million years. It seems necessary, therefore, to accelerate the accretion process, and it is not ruled out that the planet might have migrated inwards, thereby increasing the amount of material available to be captured by its gravitational field. We have seen how this process has been invoked to explain the presence of giant exoplanets close to their stars.

The second gas giant, Saturn, formed at a greater distance from the Sun, along a path where less material was available, which explains its smaller mass. The ice giants Uranus and Neptune, at 20 and 30 AU from the Sun respectively, must have been built up more slowly from the less abundant disk material available at such distances from the Sun. It is likely that their initial icy nuclei attained

Jupiter and Saturn, imaged by Voyager 1. With their rings, and satellites orbiting in their equatorial planes, these giant planets resemble miniature solar systems.

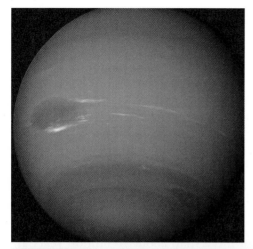

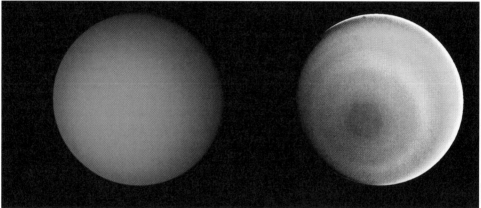

Neptune (left) and Uranus (below), imaged by Voyager 2. The two ice giants – 20 and 30 AU from the Sun – grew more slowly than the gas giants Jupiter and Saturn, which are nearer to the Sun. Uranus is shown in true colour at left, and in false colour at right.

critical mass after the protoplanetary gas had been swept away during the Sun's T Tauri phase, so that little was left to be drawn in towards them.

With their ring systems and large numbers of satellites orbiting in their equatorial planes, the giant planets resemble miniature solar systems. The collapse of the protosolar gas onto their original nuclei must have led to the formation of their own protoplanetary disks, from which the ring systems and satellites evolved. What is striking, however, is the diversity of the systems engendered by this process. Jupiter boasts four large satellites, but its rings are all but invisible; Saturn's rings are extensive, and it has many small satellites and, further out, one large one, Titan, with an atmosphere; Uranus and Neptune have very tenuous, narrow rings; and Neptune's large satellite Triton, with its thin atmosphere, is probably a captured object.

4.8 TERRESTRIAL PLANETS: DIVERGENT DESTINIES

A diversity which suggests that there may be many kinds of exoEarth... if they exist

The four inner planets of the solar system are also characterised by remarkable diversity. On the sunward side of the ice line they were formed from a quantity of refractory material only, and their accretion phase must have lasted longer than that of the giants. Computer models predict the formation, over a period ranging from a few tens of millions to perhaps 100 million years, of only a few embryonic rocky planets, of masses lower than or similar to that of Earth. Mercury, the nearest planet to the Sun, is too small and too hot to have been able to retain a stable atmosphere, and even relatively heavy gases such as carbon dioxide would, at such temperatures, have attained escape velocity. The other three terrestrial worlds have atmospheres doubtless formed partly from material delivered by meteoritic infall, including interplanetary fragments, asteroids and comets, and by outgassing.

The atmospheres of Venus, the Earth and Mars must initially have had the same global composition – carbon dioxide, water and molecular nitrogen – but each must subsequently have evolved very differently. While atmospheric pressure at the surface of Venus is currently a hundred times greater than it is on Earth, on Mars the value is only one hundredth! Venus's surface temperature of 730 K is much greater than might be expected, given its distance from the Sun. Finally, while the atmospheres of Venus and Mars are indeed composed of carbon dioxide, with a little nitrogen, the constituents of the Earth's atmosphere are molecular nitrogen and oxygen – a radically different result.

Venus, imaged by the Galileo probe in 1990, and the Earth. Venus differs from the Earth in that its water vapour has contributed to an intense greenhouse effect, raising its surface temperature to a scorching 730 K.

Mars is less massive and colder than the other terrestrial planets. Lacking sufficient mass to engender the internal energy required for the production of a dense atmosphere, It has undergone only a modest greenhouse effect.

How can this be explained? The answer most probably lies in a single molecule: water, which existed in different states according to the distance from the Sun of the three planets. Gaseous on Venus, it joined forces with carbon dioxide to create an ever-intensifying greenhouse effect, with no mechanism available to stop it. However, on Earth the temperature was such that water existed in liquid form, so the carbon dioxide was trapped in the oceans as calcium carbonate. The greenhouse effect therefore remained moderate, and Earth's surface temperature stayed relatively constant throughout its history. Due to the emergence of life on Earth, doubtless in its ocean depths, the

The greenhouse effect

The greenhouse effect is a mechanism involving the warming of the surface and lower atmosphere of a planet or a satellite. The surface, warmed through absorption of the Sun's visible light, radiates energy in the infrared. If this radiation is itself absorbed by gases in the lower atmosphere, this layer will also warm up, and the surface temperature will rise further. The mechanism is self-amplifying. Carbon dioxide and water vapour, with their strong spectral signatures in the infrared, are particularly effective greenhouse gases.

atmosphere became progressively richer in oxygen. Mars – considerably less massive and cooler than its two sister planets – had insufficient internal energy to be able to engender a dense atmosphere, and therefore experienced no strong greenhouse effect. Its atmosphere, which was probably denser in the distant past, has gradually become cooler and more rarefied, and water is now present only as ice and permafrost. It is, however, very probable that liquid water flowed across Mars in the early stages of its history; and perhaps life developed.

4.9 BETWEEN TERRESTRIALS AND GIANTS: THE MISSING PLANET?

Asteroids: embryos, prevented by Jupiter from assembling into a planet

If we consider the distances of the planets from the Sun, we can see that they are strung out according (approximately) to an exponential, empirical law: the Titius–Bode law – the formula for which is $D = 0.4 + 0.3 \, 2^n$. D is the heliocentric distance of the planet in question, and n has values of $(-\infty)$ for Mercury, from 0 to 2 for the other terrestrial planets, and then rises incrementally from 4 through the giant planets and Pluto. This relationship holds (with a 5% error) for all the planets out to Uranus, but fails in the case of Neptune (22%). Modern calculation suggests that the Titius–Bode 'law' does not represent any physical reality, but is based on mere chance.

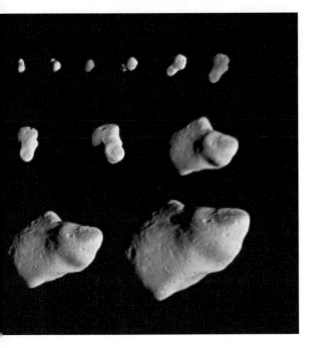

There is, however, an interesting property of this relationship: there is no planet corresponding to the value $n = 3$. What is the significance of this gap between the terrestrials and the giants? Of course, there *is* something there: a multitude of lesser bodies, the asteroids.

Most of them move within a torus-shaped region – the main asteroid belt – at distances between 2.2 and 3.4 AU from the Sun. There, at the beginning of the nineteenth century, the first asteroids were discovered: Ceres, Pallas, Vesta and

A rotation sequence of the asteroid Gaspra, imaged by the Galileo probe in 1991.

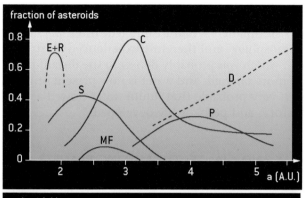

Spectral types of asteroids as a function of heliocentric distance. The silicate asteroids (S) tend to be nearer to the Sun than the carbonaceous variety (C).

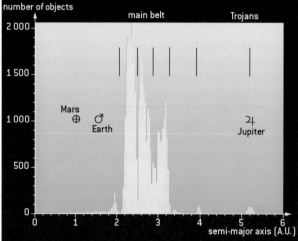

Distribution of semimajor axes of asteroids. Most of the known asteroids follow (sometimes highly) elliptical orbits, with eccentricities between 0.01 and 0.3 and inclinations not far (less than 30°) from the plane of the ecliptic.

Juno. Ceres is about 1,000 km in diameter, while all the others are smaller. More than 10,000 asteroids are now listed. Sky surveys, especially in the infrared, have largely contributed to the lengthening list of 'minor planets'.

The nature of the asteroids has been revealed to us through their visible and infrared spectra, which provides information on the composition of their surfaces. As a result of their low gravity, none of them has any atmosphere. Some are metallic in nature, some are silicaceous, and some carbonaceous. These differences seem to reflect the temperatures of their environments of origin. Globally, the densest asteroids are closest to the Sun, while the more primitive carbonaceous asteroids move through the farther regions of the main belt and beyond. Here again is evidence of the stratification of matter within the disk as a function of temperature (of distance from the Sun).

What is the origin of the asteroids? For a long time, astronomers believed that they were the remains of a 'missing planet' which, having formed between the orbits of Mars and Jupiter, could not survive because of the gravitational effects caused by its mighty neighbour. Today, due to the development of computer

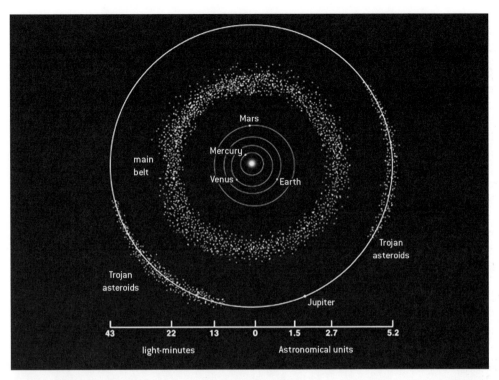

The main asteroid belt – where the greatest concentration of asteroidal orbits is found – lies between the orbits of Mars and Jupiter (2–3.5 AU from the Sun). Beyond this belt orbit the Trojan groups, following Jupiter's path; and further out still, the Centaurs move between the orbits of Jupiter and Neptune.

models simulating the formation of celestial bodies, it is thought that the mythical planet never existed – again because of the presence of Jupiter. Most of the planetary embryos became members of the main asteroid belt.

Remember that there are other 'minor planets' which follow other orbits. Of most interest to us are the 'Earth-grazers', the orbits of which are close to that of the Earth. Although the probability of a collision is very small, we cannot discount this kind of hazard completely. The history of the Earth offers evidence of such events in the past, with consequent profound effects upon its ecosystem. Certainly, the impact of an object about 10 km across was, for the most part, responsible for the demise of the dinosaurs at the Cretaceous–Tertiary boundary (K–T boundary) 65 million years ago.

4.10 THE ORIGIN OF COMETS: THE KUIPER BELT AND THE OORT CLOUD

Comets and trans-Neptunian objects of the solar system. . . and perhaps of other systems

Continuing down our list of solar system objects we come to the comets: small, but by no means little known. Comprising a nucleus of ice and rock (usually no more than 10 km across), these wanderers in space follow very elliptical orbits, usually far from the Sun and the Earth, but when they approach us they can appear as spectacular objects. As the ices of the nucleus sublimate, gas and dust are ejected, reflecting sunlight and producing long luminous trains. Since ancient times, comets have inspired fear and superstitious awe. Not until the eighteenth century was the nature and periodicity of comets identified, by English astronomer Edmond Halley, who is chiefly remembered for his prediction of the return, in 1758, of the comet which now bears his name.

Composed mostly of water ice, comets are the remnants of the first embryonic bodies formed in the outer solar system. In what environment did they originate? Studies of their orbits provide accurate information as to their provenance. The currently accepted scenario has comets forming at a few tens of AU from the Sun, beyond the orbit of Uranus. Some of them (probably the

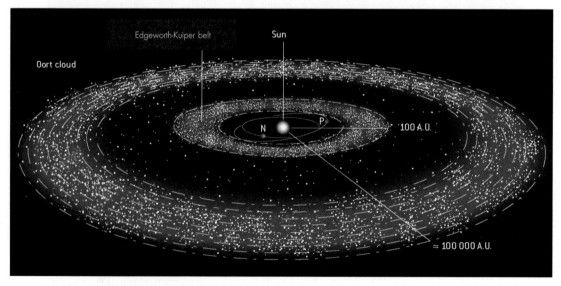

Where do comets come from? The two reservoirs of comets in the solar system are the Oort Cloud, between 10^4 and 10^5 AU from the Sun, and the Edgeworth-Kuiper Belt, between 30 and 100 AU. This diagram (not to scale) shows a section through these reservoirs along the plane of the ecliptic. The Edgeworth–Kuiper Belt is situated more or less in this plane, but the Oort Cloud is somewhat shell-like in form, extending to high heliocentric latitudes.

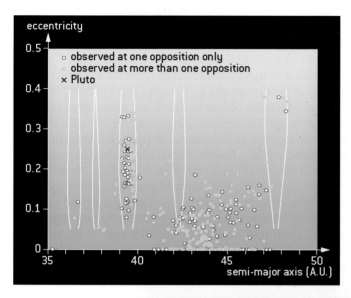

Eccentricities of trans-Neptunian objects as a function of the semimajor axes of their orbits. The dots represent the positions of objects with well-known orbits. The different zones of resonance (4:3, 7:5, 3:2...) are shown. Objects which, like Pluto, are localised at around 39 AU, are called Plutinos. They are in a 3:2 resonance with Neptune (their respective periods are in the ratio 3:2).

The nucleus of Halley's comet, imaged by the Giotto probe in 1986. The lighter parts correspond to sites of gas and dust emissions.

nearest to the Sun) were ejected from the main solar system by gravitational effects due to the giant planets, especially Jupiter. These comets now populate the Oort Cloud – a spherical halo of bodies situated between 40,000 and 100,000 AU from the Sun. At this distance some of them already lie at a third of the distance to the next star! External perturbations sometimes send comets inward from the Oort Cloud, towards the vicinity of Earth. There are also comets which formed beyond the orbit of Neptune. These form part of a recently discovered population of objects collectively known as the Kuiper Belt, which contains the remnants of planetesimals formed within the protoplanetary disk beyond the orbits of Uranus and Neptune. It lies in the ecliptic plane, between 30 and 100 AU from the Sun. Its existence was predicted in the mid-twentieth century, as a result of theoretical studies by the astronomers Kenneth Edgeworth and Gerard Kuiper. Not until the 1990s did it become technically possible to detect Kuiper Belt objects. About 1,000 trans-Neptunian objects are now known. It is probable that they have large icy nuclei, and the largest of them have diameters of several hundred kilometres. It now seems that Pluto – a body previously classed as a planet which has long intrigued astronomers with its unusual nature and eccentric orbit – is merely one of the largest members of this new class of object. The existence of the Kuiper Belt has taken its place within the framework of exoplanetary research, and the region may well have equivalents in other protoplanetary disks or in the outer environments of other stars. It is possible that an excess of water vapour around the evolved star W Hya is evidence of the volatilisation of its Kuiper Belt, heated by the star as it evolved into a red giant.

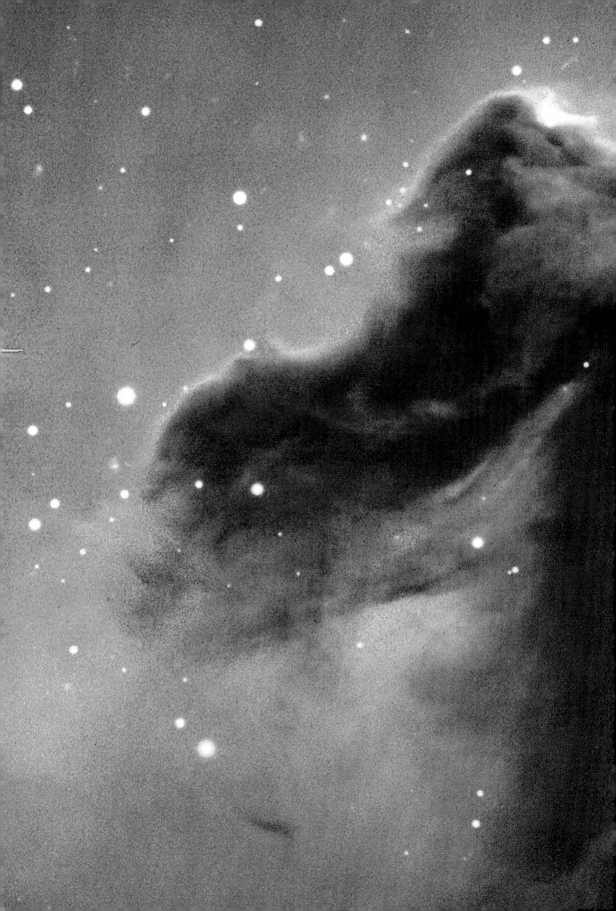

5
The formation of planetary systems

The exoplanets so far discovered seem to have been formed differently from the planets of the solar system. To explain the 'hot Jupiters' seen in such large numbers in extrasolar systems, we must try to imagine a process involving the migration of planets from the outer reaches of the disk towards the central star.

The Horsehead Nebula – a star-forming region in the constellation of Orion ▸

5.1 STAR-BIRTH

From disks to jets, through accretion and ejection: a birth process lasting a million years

To understand how planets are formed we must delve into the formation of the stars. The two processes are intimately related: planets are born at the same time as stars.

How do stars come to be? They are continuously forming within our galaxy, and it is estimated that every year a new Sun begins its career. It all begins with an interstellar cloud – a vast region of gas (hydrogen and helium) and dust consisting of silicates and carbon aggregates, often coated with ice. Such a cloud is immense, measuring tens of light-years across, but relatively tenuous. Each cubic centimetre contains a few dozen atoms. Interstellar clouds are cold, at 10–20 degrees above absolute zero. Within this mighty cloud there are regions of above-average density. The smallest of these knots of matter are just a few thousand AU across. They are mostly stable, but they can become destabilised, for example, by a nearby supernova explosion. Gravity then takes over, and the knot begins to condense as a result of its own mass.

Let us follow the evolution of one of these condensed regions, as it becomes a star and its protoplanetary disk. The matter falls in upon itself, and at its centre

Star-forming regions in the Eagle Nebula (M16). This image was obtained by the Hubble Space Telescope's wide-field camera (WFPC2) in 1995. The 'pillars' are columns of relatively cool molecular hydrogen and dust on the outskirts of a dense interstellar cloud.

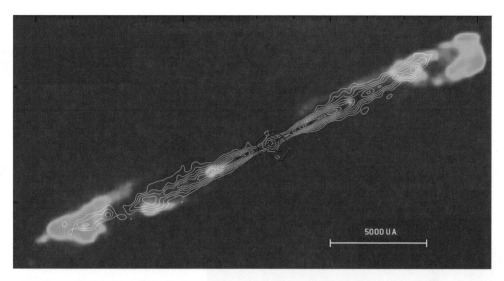

5000 U A

The gravitational collapse of a molecular cloud and the formation of a protostar are systematically associated with jets of material issuing from regions close to a star. This map shows such a molecular jet flowing out from the protostar HH 211. The CO emission (white lines) is traced out on this false-colour image showing molecular hydrogen. The red lines represent a flattened concentration of dust around the protostar.

becomes denser and denser. The temperature rises. Not more than a few million years pass, and the temperature necessary to trigger hydrogen fusion is attained: 3,000,000°. A new star is being born. Around this protostar, a disk of gas and dust forms. Now, even if the knot of matter had been more or less spherical to begin with, the slightest rotational tendency within it will lead to the outer parts of the collapsing mass, forming a flattened disk. In the direction perpendicular to the axis of rotation, centrifugal force opposes gravity. Within this protoplanetary disk, planets will form. Their orbits will be almost coplanar, as has occurred in the solar system (see p. 43).

This disk surrounding a young star, with matter from the original knot continuing to fall inwards, is an accretion disk. Another phenomenon occurs. Observation has revealed that young stars systematically emit vast jets of matter at right angles to the accretion disk. This behaviour is still not well understood, but it seems to be universal. Simultaneously, matter is falling towards the star (accretion) and an ejection process is occurring. Between accretion and ejection, planets are forming within the disk.

Research into star formation is one of the most challenging aspects of modern astronomy. Monitoring it is not easy, as young stars are enveloped in a cocoon of gas and dust, making it difficult to observe them in the infrared and radio domains and impossible to see in the visible. On the theoretical side there are similar difficulties, as no model yet exists to describe the process of star formation in its entirety, as matter passes from a state where there are just a few tens of particles per cubic centimetre to arrive at a density several billions of billions of times greater.

5.2 INSIDE PROTOPLANETARY DISKS

Are planets forming deep within the disk?

Over the past decade astronomers have gradually begun to reveal the mechanisms within protoplanetary disks. Visual observation is not easy, as the disks are generally opaque at visible-light wavelengths. We therefore have little information about the processes in these regions, but we are able to glean some knowledge from studies with radio telescopes in the millimetric and centimetric domains, and with interferometers, such as the VLTI, in the infrared. Space platforms are also used, working in the far infrared. However, the spatial resolution achievable with these types of observation is rarely good enough to secure images of disks.

Most protostars seem to have their accompanying disks. Early on, they were observed around smaller, T Tauri-type protostars (see p. 71), analogous to the Sun when it was a million years old. It is now known that more massive stars also have them. Examples of these disks seen edge-on, as imaged by the Hubble Space Telescope, reveal their dimensions and shapes. Their diameters may be greater than several hundred AU, and they seem thicker in their outer regions. Radio observations have confirmed

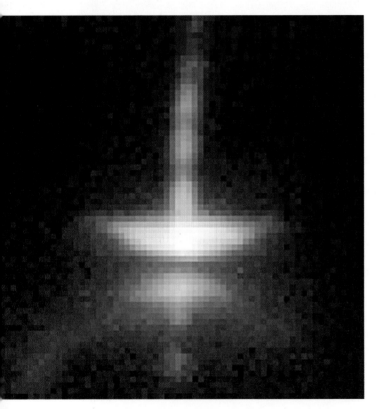

The protostar HH 30, imaged by the Hubble Space Telescope. HH 30 is surrounded by a thin, dark disk, and is emitting powerful jets of gas (seen in green). The central star is hidden, but its light is reflected from the upper and lower surfaces of the disk, the bright areas giving the nebulous material the appearance of a yo-yo.

The VLTI interferometer at Paranal Observatory, Chile. This photograph, taken in March 2005, shows the four 8.2-metre telescopes of the ESO's Very Large Telescope (VLT). The VLTI is able to link these instruments, and three auxiliary telescopes, together as an interferometer.

that these disks are rotating around their stars. They are of low mass, between 0.001 and 0.1 of a solar mass, but this is quite sufficient for the formation of a few giant planets.

What do protoplanetary disks contain? First of all, gas – obviously hydrogen, as this is the most abundant element in the Universe, although it cannot be detected directly. Simple molecules, such as carbon monoxide, and more complex molecules, such as methanol, are also present. These are perhaps the early building blocks of prebiotic chemistry. The density of the gas, as well as the temperature, increases towards the centre. The very opacity of the disks shows that dust is present in large amounts. Observations in radio and infrared wavelengths (most recently by the Spitzer Space Telescope) show that these dust grains are not the same as those in the initial interstellar cloud. In particular, the infrared spectra of certain disks suggest that they contain grains of centimetric size, while interstellar dust grains are scarcely larger than 1 μm. Since these observations cannot detect bodies larger than centimetric size, it is probable that these 'large' grains indicate the presence of planetesimals. As for the presence of planets, these cannot be directly observed, and young stars are far too variable for velocimetric studies to be viable.

Disks are bathed in the radiation from their central stars and are subject to the vicissitudes of their eruptions and stellar winds, which are far more violent than those of Sun-like stars. Young, nearby stars contribute their ultraviolet radiation, and the disks (and especially the gas within them) therefore have limited lifetimes. Almost no stars more than a few million years old are known to retain their gaseous disks. And with the gas gone, giant planets cannot easily be formed; so in just a few million years – perhaps 10 million years at the most – planetary systems have to be built.

5.3 PLANETARY EMBRYOS

Planets form as dust coagulates and accumulates more matter

Thirteen orders of size separate a grain of interstellar dust less than 1 μm across from a terrestrial planet. A long road stretches between them – a road which has to be travelled in less than 10 million years. Dust collects into 'grains' within the newly formed protoplanetary disk. These grains settle into the central plane of the disk to form a thin layer, much thinner than the gas disk, the volume of which is maintained by its own pressure. The density is high enough for the grains to be driven by the gas into frequent mutual collisions, and they adhere to each other electrostatically. So, by a process of coagulation, grains a few centimetres across are formed, and have been observed in disks around protostars. Computer models reveal that this initial phase probably lasts a few thousand years.

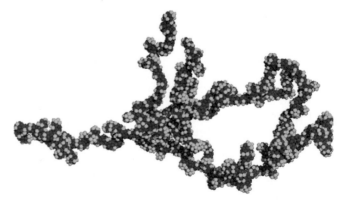

The next stage – the building of planetesimals with diameters of a few kilometres – is less well understood. Do the planetesimals form as a

Fractal structure of an interstellar dust grain. A 'grain' of interstellar dust probably has a complex structure as depicted in this simulation. This fractal arrangement doubtless facilitates the agglomeration of dust grains when they collide.

The cratered surface of the Moon bears evidence of enormous impacts: protoplanets collide with each other, resulting in the formation of a number of more massive bodies.

result of gravitational instability within a disk that has become very dense? One advantage with this process is that it is rapid. The drawback, however, is that it is easily affected by any residual turbulence within the gas disk. So, while planetesimals are not yet of kilometric size, their dynamic is determined by the movements of the gas. Do the grains become planetesimals by collision and coagulation, as in the preceding phase? It is probable that the two processes, gravitational instability and coagulation, work together. More observations are needed for us to understand the effects of turbulence within the gas of protoplanetary disks.

By whatever route, we have now reached the stage of planetesimals – billions of bodies a few kilometres across, orbiting the protostar within a disk which still retains its gas. At this stage, gravitational interactions between planetesimals become important. Now begins the great celestial billiard game, with bodies attracting each other and colliding. In this game there will be winners. Some of the planetesimals will begin to grow larger than others and will attract smaller bodies towards them, 'cleaning out' wide circular regions around the star as they do so. As this process of accelerated growth continues, a certain number of bodies of very respectable mass will form over a few tens of thousands of years, proceeding in almost circular orbits.

Now comes the era of mighty impacts. One of the protoplanets, gravitationally perturbed by neighbouring bodies, will move out of its circular orbit and collide with another protoplanet. This process will generally lead to the destruction of the smaller of the two bodies. Hundreds of millions of years

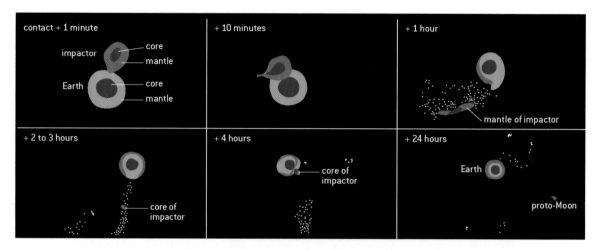

Stages in the formation of the Moon as a result of a gigantic impact, upon the protoEarth, of another body. (After Cameron.)

later, this natural selection will leave only a few survivors in more or less stable orbits – and a planetary system is born.

Analytical calculation and numerical simulations describing the formation of a solar system are not easily carried out, and the processes to be modelled are complex and in some cases little understood. Furthermore, achieving the initial conditions is extremely difficult. If one of the hundred or so Moon-sized bodies in our simulation is just a little out of place, we might end up with two instead of five terrestrial planets!

5.4 RECIPES FOR A GIANT PLANET

A heart of ice and rock, and a dense atmosphere

Dust coagulates, planetesimals accumulate: a good enough recipe for a terrestrial planet, but what are the ingredients for a giant? Researchers are divided on this question. Most of them favour an explanation involving accretion onto a solid core. This is a mechanism that has probably prevailed with the giants of the solar system, but a small minority actively promotes a scenario of instability with the gaseous disk.

Let us first examine the more commonly proposed mechanism. An initial core of rock and ice has to form, with a mass a few times greater than that of the Earth. But this cannot occur just anywhere within the protoplanetary disk. Because a protoplanet grows by sweeping up whatever it encounters in a thin ring within the disk, the maximum mass of that protoplanet is that of the ring itself. In the vicinity of the star – for example, at a few hundredths of an AU –

Two types of giant planet, according to their distances from the Sun. Jupiter (left) is essentially made of protosolar gas, while Neptune consists mostly of ice.

there is no dust, as it has been vaporised by temperatures greater than 2,000 K, and there is probably no gas. So, no planets can possibly form there. A little further away from the star, dust is present, but the disk surface 'swept' by the protoplanet is small, and terrestrial planets can form. At a few AU from the star the surface 'swept' is greater, and the situation is more favourable. Beyond the ice line, dust particles will be covered in ice.

When the mass of the core exceeds about 10–12 Earth masses it becomes capable of drawing in all the surrounding gas, and of retaining it, even if subsequently struck by planetesimals. The protoplanet, enveloping itself in a thick hydrogen–helium atmosphere, rapidly becomes a giant – as long as there is still a disk of gas around the star! This implies that the process has to take place within just a few million years. Therein lies the difficulty. It takes a long time for a core ten times the mass of the Earth to come into being. Everything depends on the initial density of the disk. If that density is just sufficient to produce, for example, the eight planets of the solar system, tens of millions or even hundreds of millions of years are required. At slightly greater densities, as have been observed in some disks, a core equivalent to that of Jupiter or Saturn has a reasonable time in which to form. This scenario explains why Uranus and Neptune, both very near the critical mass, possess only relatively thin gaseous envelopes, and are ice giants rather than gas giants. They no doubt attained the critical mass at too late a stage, when no gas remained nearby.

What do the minority voices say? They remind us that there is no absolute certainty that Jupiter and Saturn actually have rocky cores. Citing the time-scale

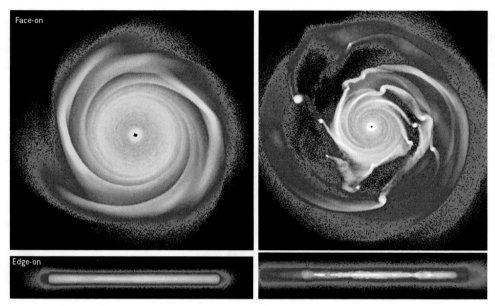

Computer simulations of the formation of giant planets in a disk of gas, as a result of gravitational instability. These images show the evolution of the disk at t = 160 years and t = 350 years. Clumps are forming rapidly. Their masses and heliocentric distances resemble those of giant planets, but their composition does not. The radius of the disk is 20 AU. (Simulation by L. Mayer, J. Wadsley, T. Quinn, J. Stadel (University of Zurich), McMaster (Canada) and Washington, 2003.)

problem – in that the formation of such cores would be a very long process if the density of the disk were insufficient – they propose that a small knot, denser than the average material within the gaseous disk, collapses in upon itself to form a gas giant. Terrestrial and giant planets, they say, are produced in totally different ways, with the formation of the giants resembling that of the stars. There is, however, a small problem in that this avenue of approach points towards the composition of gas giants being much the same as that of the Sun, although they are richer in heavy elements.

5.5 MIGRATING PLANETS

Hot Jupiters to the fore again

A scenario within which giant planets form at several AU from their star is not a completely comfortable one for researchers. How do we explain those hot or even scorching Jupiters – those Pegasids which have been found by the dozen? What of those large planets dominating regions at just a few hundredths of an

Computer simulations of the formation of a gap in a protoplanetary disk by a planet of 1 M_J. The interaction of the planet and the surrounding gas causes movements in the gas to depart from the circular, and lines of flow are modified. By the same token, the motion of the planet around the star is also modified. (After Angelo, Henning and Kley.)

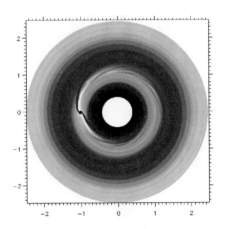

Gaps in protoplanetary disks should be observable with future instruments such as the ALMA interferometer, currently under construction in northern Chile. Here, simulations suggest what ALMA might 'see' if the disk is viewed face-on (left), at a tilt (centre), or almost edge-on (right). (After Wolf, Gueth, Henning and Kley, *Ap. J.*, **566**, 97 (2002).)

AU: for example, 51 Pegasi b, the first exoplanet to be discovered? Also, the low density of HD 209458 b implies that we are here dealing with gaseous and not terrestrial planets, and we would be hard put to justify the existence of a completely rocky planet in this region. How could this be possible, in a zone where even the most refractory dust particles would be vaporised? The only answer seems to be that the planet was formed in the outer part of the disk, and then migrated inwards.

It is worth remembering that theoreticians had predicted this unusual phenomenon of migration as early as 1979, well before the discovery of 51 Pegasi b, but were not easily able to explain why it had not happened in our solar system! Migration seems inevitable for any planet formed while the proto-planetary disk is still present: the matter in the disk, and the planets, fall inexorably towards the protostar. The driving force behind this phenomenon is the interaction between the disk and the planet. Around the planet there is a region where its own gravitational effects outweigh those of the star. In this region the force of gravity varies with the distance from the planet, creating tidal forces like those acting upon the Moon as it moves through the Earth's

An empty gap created by a planet (in red) within a gaseous disk. (Simulation by G. Bryden, Jet Propulsion Laboratory, Pasadena.)

gravitational field. These forces cause concentrations of matter to form within the disk – one between the star and the planet, and another beyond the planet, spiralling outwards as the disk rotates. The motion of the gas is no longer circular, and currents within it are modified. The presence of the planet has therefore modified the rotation of the gas, but the planet's own motion around the star has also been altered.

The net result of these subtle interactions between disk and planet is striking: the planet pushes away gas on either side, often creating an empty gap in the disk. At the same time, both planet and disk fall inwards towards the star. These migrations are classified into two types. A Type I migration occurs when no gap is opened, but the planet still falls rapidly towards the star over a period of about 10,000 years. Planets involved are mostly less massive than Jupiter. Type II migrations involve massive planets opening gaps and falling at the same rate as those gaps. In this case, the planet may take several hundred thousand years to reach the star.

There are many other reasons why a planet might migrate, including interaction with another giant planet, interaction with a distant stellar companion of the central star, forces of drag experienced as the planet moves though a disk of planetesimals, and so on. In a sense, it is no surprise that so many hot Jupiters have been found!

5.6 SURVIVAL STRATEGIES

Do migrant planets settle instead of falling into stars?

It should come as no surprise that certain planets are eventually swallowed by their stars, as the same fate awaits comets within the solar system. This phenomenon has also been put forward to explain why stars with planets contain more heavy elements than other stars. Migrations take remarkably little time: 100,000 years, or even less for Type I events! This seems to suggest that planets are no sooner born than they are devoured. How is it, then, that we are discovering planets by the dozen? For migration to occur, the prerequisite is the existence of a disk, whether it is gaseous or consists of planetesimals. As soon as the disk dissipates, migration ceases. We can picture an evolution, therefore, during which several successive generations of planets are born within the disk and are swallowed up, while only the last brood, formed shortly before the dissipation of the disk, will survive.

Evolution of a disk of material containing two protoplanets.
These are as massive as Jupiter, but the semimajor axis of one is twice that of the other. They create two gaps, separated by a ring of gas (upper image), which then thins (centre) and finally disappears (lower). (Simulation by G. Bryden, Jet Propulsion Laboratory, Pasadena.)

The rings of Saturn, imaged by the Cassini spacecraft.

We can also envisage other ways in which the headlong fall of a planet might be stopped just before it tumbles into the star. It has been stated that no planets are known with orbital periods shorter than 1.2 days. The protoplanetary disk probably stops short of the star, held at bay by its stellar wind or by interactions between the gas and the star's magnetic field. The mechanism driving the migration would cease to operate within the gap thus created. We might also theorise that the planet could, at such a short distance from the star, be subjected to an intense gravitational field which causes it to lose mass, or to be deformed by tidal effects. Either of these scenarios would break its fall. All these phenomena are complex, and difficult to model. To understand them better it will be necessary to observe the inner regions of disks in detail. This will be the task of a new generation of instruments: radio and infrared interferometers.

The situation is even more complex, for in reality it is not just a single-planet disk that we have to consider, but one containing several planets. This reflects the general case, and changes everything: the few attempts at a digital description of the problem suggest that planets can even migrate outwards! Perhaps this is what happened, within our solar system, to Uranus and Neptune. These two ice giants no doubt formed nearer to the region of the gas giants, and then moved away from the Sun, drawing along in their wakes a collection of planetesimals. These planetesimals may well now populate the Kuiper Belt – the region beyond Neptune containing billions of icy bodies, some as big as 1,000 km across. Such outward migrations towards the periphery of the system might also explain the presence of giant planets at tens of AU from their stars, as in the case of the planet of 5 Jupiter masses orbiting at 55 AU from the brown dwarf star 2M 1207.

An example of the extreme complexity of a system, as a result of gravity alone, is provided by the rings and satellites of Saturn; and the presence of gas in a protoplanetary disk does nothing to simplify matters. So, research in this field proceeds apace. The questions still unanswered includes: what of the survival of any terrestrial planets? If they are drawn into the zone of influence of a giant, they too may be swallowed by their star.

5.7 PLANETS AROUND PULSARS

Two known systems with very different stories

Among the forty or so millisecond pulsars so far identified within our Galaxy, only two are known to possess planets. On the face of it, this rate of detection seems no better nor worse than in the case of other types of star studied using the velocimetric method. Remember, though, that the pulsar timing method can detect bodies as small as an asteroid, so if no planet is found the result is far more significant than a negative result from radial velocity studies. Pulsars with planets are indeed rare phenomena.

The two known examples are very different from one another. PSR 1257+12 has three terrestrial-sized planets – two of them of 4.3 and 3.9 Earth masses. They all orbit more closely to the star than Mercury to the Sun. Two of them are in a 3:2 resonance (as one completes exactly three revolutions about the pulsar, the other completes two orbits). Several theories have been advanced to explain the presence of these planets. If the progenitor star had any planets, they could not have survived when it exploded as the supernova of which the pulsar is the remnant. However, if there had been a protoplanetary disk or a sufficiently massive disk of debris several hundred times the mass of the Earth, it may have survived, its remains becoming planets. There is another possibility. If the pulsar had a stellar companion that was dislocated by tidal forces, this might have given rise to a disk within which planets could be formed. If the pulsar was born out of the fusion of two white dwarf stars, a disk might also have resulted. White dwarfs represent the last stage of fairly low-mass stars like the Sun. All these processes are possible, but they do not happen every day!

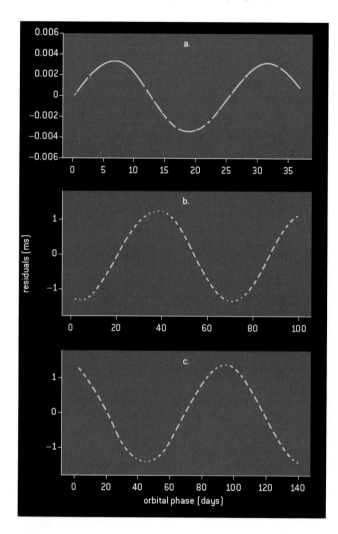

Pulsar timing. PSR 1257+12 has three planets of terrestrial mass, all orbiting closer to the pulsar than the distance of Mercury from the Sun. Two of these planets are in a 3:2 resonance. The figure shows the sinusoidal variations in the arrival times of the pulsar's signal, due to each of these planets. (After Wolszczan, *Science*, **294**, 5158 (1994).)

PSR 1620–26 is a very different object. In this system there is a planet of Jupiter-type mass (2.5 M_J) orbiting a pair of stars – one a white dwarf and the other a neutron star (the pulsar). This system is located within the globular cluster M4. One suggested reason for the presence of this planet is truly surprising: it may have been 'borrowed' from another star! Within globular clusters, stars are close enough together for frequent interactions to occur. In this scenario a star accompanied by a planet passes near a pair of neutron stars, and changes places with one of the neutron stars. The planet, still with its original star, assumes a new orbit around the new pairing, which now consists of a neutron star and an ordinary star which later evolves into a white dwarf. This is certainly a configuration which cannot be very common, and is an interesting problem in celestial mechanics.

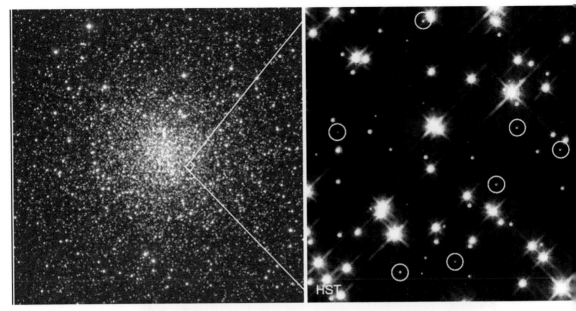

The globular cluster M4, imaged by the Hubble Space Telescope in 1995 (left). A close-up (right) shows an area 0.63 light-years across. The HST detected 75 white dwarf stars (circled) in this region, and in the whole cluster there must be some 40,000 of them. Pulsar PSR 1620–26, also in this cluster, has a white dwarf companion and a Jupiter-sized planet probably 'borrowed' from another star.

PSR 1620–26 b is one of the rare examples of a planet in an environment deficient in heavy elements. As we have seen, searching for planets by the radial velocity method is more effective if the target stars have a heavy-element content greater than that of the Sun. The solar system is an obvious exception to this rule; and PSR 1620–26 is another. The globular cluster harbouring it consists of stars which are among the most deficient in heavy elements. Here we have yet another mystery of planet formation.

5.8 STABILITY IN PLANETARY SYSTEMS

What do extrasolar systems hold in store for us?

How unstable might extrasolar planetary systems be? Considering the variety of possible dynamical phenomena in a system – such as collisions, and planet migrations either inwards or outwards – this is, of course, a question which must arise.

We know that a certain degree of chaos exists within the solar system. But is this typical of other systems? We know of twenty-one 'systems' of two or more

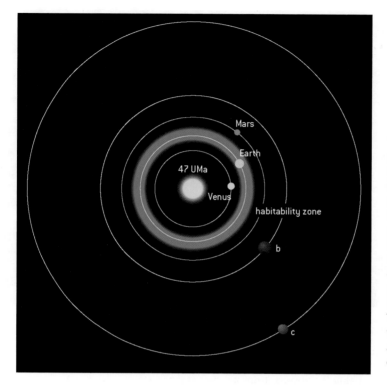

The habitability zone of the double-planet system 47 Ursae Majoris. This zone lies within a thin ring around the star, within which liquid water might be able to exist long enough on a planet's surface for life to emerge there. The orbits of Venus, the Earth and Mars are shown for comparison.

planets, and the number is fast increasing. Massive worlds and highly eccentric orbits – the conditions for the possible ejection of one of the planets. Specialists in planetary dynamics have been investigating this problem, perhaps optimistically, as the parameters of these systems are not well known. Not only is the well-known inclination factor likely to muddy the waters when planetary masses are determined, but we do not know the possible relative orbital inclinations of the planets involved. Even if they are more or less coplanar – which is probable when we consider the layout of the solar system – there is no reason to think that they are exactly so. If they are not, then this has ramifications for the evolution of the system. Furthermore, the inventory of planets in each of these systems is by no means complete. There may well be, in remote orbits, massive worlds as yet undetected.

In order to study the stability of a planetary system we run its evolution using the orbital parameters deduced by observation, allowing for small variations compatible with observational error. In certain cases the system will exhibit chaotic behaviour, while in others a planet will be finally ejected – or nothing much may happen. Calculations show that most of these planetary systems are perilously close to zones of instability. Systems with a 2:1 resonance (the period of an outer planet being twice as long as that of an inner planet) are particularly unstable. Although the inner planet is unlikely to be ejected, the outer one runs a high risk of such a fate. A notable exception is that of the system of pulsar PSR

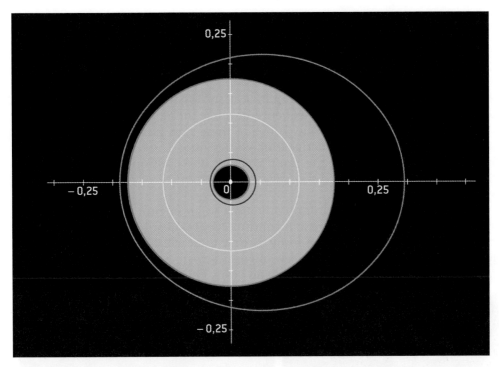

The quadruple-planet system of 55 Cancri. This star has at least four planets. The orbits of the three outer planets are shown here. Dynamical studies reveal that if an Earth-sized planet moves within the habitability zone its orbit is certainly stable.

1257+12, with three terrestrial-type planets. Their orbital parameters, masses and inclinations are comparatively well known, and the system seems well settled and unlikely to experience any modification for at least the next billion years.

Another, and trickier, question is this. If any of these systems were to contain a planet of the same mass as the Earth – an exoEarth within the habitability zone – would it be likely to remain there? As will be seen later, this zone exists as a narrow ring around the star, within which ring water may persist in liquid form on a planetary surface for a period long enough for life to develop. The situation varies from one planetary system to another. In the two-planet system of 47 Ursae Majoris, no terrestrial planet can survive in the region about 1 AU from the star. However, the situation of the four planets of 55 Cancri seems more promising. To remain in the habitability zone for a few million years without being ejected seems like a good start for a terrestrial planet; but who is to say that this makes it a 'habitable' planet?

5.9 DISKS OF DEBRIS

Dusty disks: evidence of the presence of planets and comets

Protoplanetary disks last only a few million years. We hardly expect, therefore, to see them around mature, main-sequence stars like the Sun, quietly burning away their hydrogen. But... about one in five of these stars possesses a dust disk, its presence betrayed by its distinctive infrared signature. Dust – but little or no gas, which was all dispersed in the star's younger days. Astronomers believe that these dust disks (or debris disks) suggest the existence of a mature planetary system orbiting the star.

Dust particles have a very limited lifetime in an environment like the solar system, as a particle orbiting a star is continuously bombarded by photons. In the case of small dust particles the result is that they will be drawn in, spiralling towards the star, and a particle a few microns across, at a distance of 1 AU, will fall into the Sun after only a few thousand years. This is a very efficient 'cleaning'

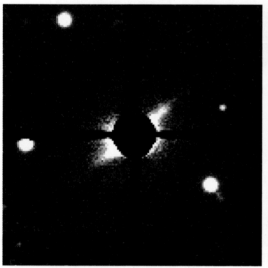

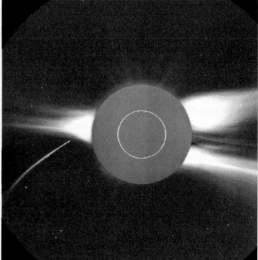

A visible-light image of a disk around the red dwarf star AU Mic, discovered in 2004 by P. Kalas with the telescope of the University of Hawaii. The radius of the disk is about 210 AU, and is very similar to the disk of β Pictoris, which is twice as far away. The absence of material in the central part of the disk (within 17 AU from the star), and its irregular structure, suggest that planets may already have formed there.

Comet SOHO 6 falling into the Sun. This image from the Solar and Heliospheric Observatory (SOHO) shows the outer atmosphere of the Sun (the corona) and comet SOHO 6 (at lower left). The Sun (at the centre) has been masked using the technique called coronography. The luminous plumes around the Sun are jets of charged particles (the solar wind) in the corona.

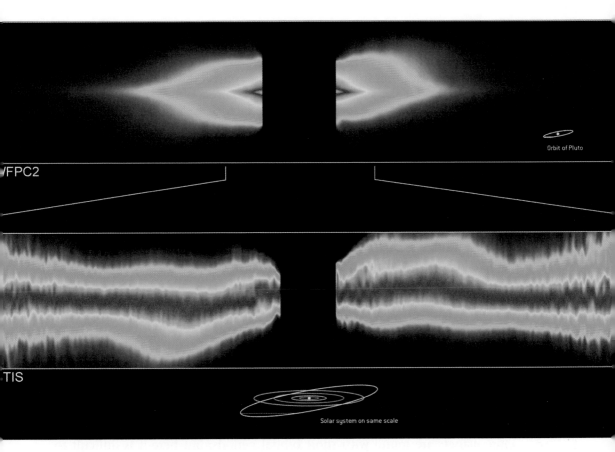

Orbit of Pluto

/FPC2

TIS

Solar system on same scale

The disk of the star β Pictoris, imaged by the Hubble Space Telescope using the coronographic method. The perturbed and somewhat 'misty' appearance of the disk may be the result of the presence within it of one or more massive planets.

process. So, if there is dust around stars tens of millions of years old, there must somewhere be a source of supply. Could it be produced by collisions of asteroid-sized bodies, or outgassing by comets? Whatever the explanation, the existence of debris disks must involve the presence of more or less 'completed' planetary systems. There is certainly plenty of dust within our solar system, of course – much of it left by the wakes of comets.

The first of these 'infrared signatory' stars was Vega; but the prototype will always be the dust disk around β Pictoris (the second brightest star in the constellation of Pictor, the Painter). The first image of the β Pictoris debris disk was secured in 1984, with the light of the star masked using the technique of coronography. Since then, the system has been studied in some detail. The disk appears to be asymmetrical – perhaps because of the existence of a giant planet at about 10 AU from the star. In other debris disks, localised concentrations are observed which may indicate the paths of planets.

There is little gas in these disks, but what there is can be very interesting. β Pictoris is an example. Its spectrum shows lines varying considerably in the course of just a few hours. The only plausible explanation is to ascribe these variations to gas escaping from the surfaces of comets during close encounters with the star! In our solar system we see comets approaching so close to the Sun that they fall into it. This is happening in the β Pictoris system hundreds of times a year. What is it that is perturbing these comets to such an extent that they are falling into their star? The most probable answer is: one or more giant planets. Here, then, is another star with a planetary system, even if there is no direct proof of its existence.

5.10 THE SOLAR SYSTEM: AN EXCEPTION?

What have we learned about planetary systems?

Let us now summarise what we have learned about planetary systems. How well do we understand them? We return to those 200 exoplanets already discovered. What of their statistical properties?

Low-mass exoplanets are the more numerous, favouring a model whereby planets form around a rocky core, itself the result of the agglomeration of planetesimals. Planets also seem to be more often found orbiting stars rich in heavy elements, which strongly suggests that these systems were formed from knots of interstellar matter laden with dust and heavy elements, increasing the probability of resultant massive protoplanetary disks.

More planets are found with short orbital periods; but here it is difficult to distinguish between the effects of observational bias (short-period planets are easier to detect) and the consequences of migration phenomena, which models suggest are omnipresent, and omnipotent! None of these planets has a period shorter than 1 day. In spite of the many explanations proposed to explain the cessation of planetary migrations at a short distance from the star, none has been proved. There must, however, be some effective mechanism at work, or we would not be detecting so many planets.

As far as the shapes of orbits are concerned, most models show planets attaining their final masses in orbits of low eccentricity, and markedly elliptical orbits therefore pose a problem. They are, though, a natural consequence of interactions between a number of giant planets. For systems with just one giant we can surmise that there were originally others but that they have all been ejected.

Finally, for every planetary system we can find good reasons why it is as it is. In the infinite variety of observed systems, every one of the complex phenomena governing the formation of planetary systems may find its place. There are still plenty of unanswered questions, however – the first being why one star should be accompanied by its retinue of planets, while others have none. A general description of the reasons for this would also invite another question. Why do

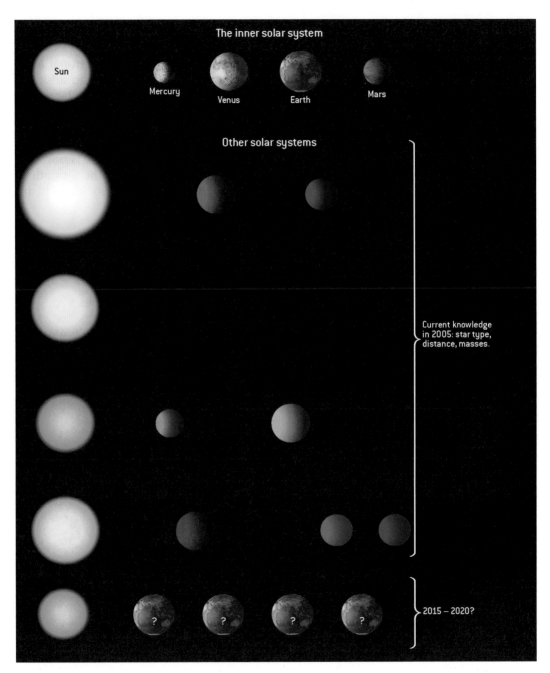

What do we know about planetary systems? We know the solar system (top) in detail, and we know that some stars have giant planetary companions with orbital characteristics and minimum masses that we can deduce. For more extensive studies of these planets we shall have to wait for the new instruments of the next decade to become available.

some stars evolve a complete system, while others show arrested development at the stage of a debris disk?

In this array of different systems the solar system is not as exceptional as it may at first seem. A planet like Jupiter would not be out of place in another system, and we know of other examples of massive planets in near-circular orbits a few AU from their stars. Whether there are 'Saturns' around other stars is something we are not yet in a position to confirm. More difficult still is the question of the excess of heavy elements in stars with planets; even so, we know of several of these stars less rich than the Sun in such elements. There is nothing in current observational findings which leads us to conclude that the formation of the Sun and its family of planets – one of which is home to life – is an exceptional phenomenon.

5.11 THE SPECTRA OF EXOPLANETS

Spectroscopy reveals the composition of exoplanets' atmospheres

Most of the information we have about the planets of the solar system is gleaned from studies of their electromagnetic radiations; that is, from the analysis of radiation as a function of wavelength, from ultraviolet through visible light to the far infrared.

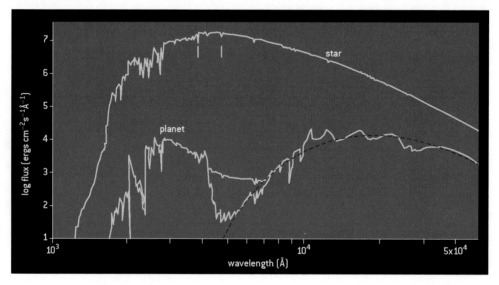

Models of the relative flux from a star and a planet (τ Böo, τ Böo b) in visible light and infrared. Above, the spectrum of the star τ Böo, and below, that of τ Böo b with (lower curve) and without silicate dust. The red curve corresponds to a blackbody at a temperature of 1,580 K. (After S. Seager and D.D. Sasselov, *Ap. J. Lett.*, **502**, 157 (1998).)

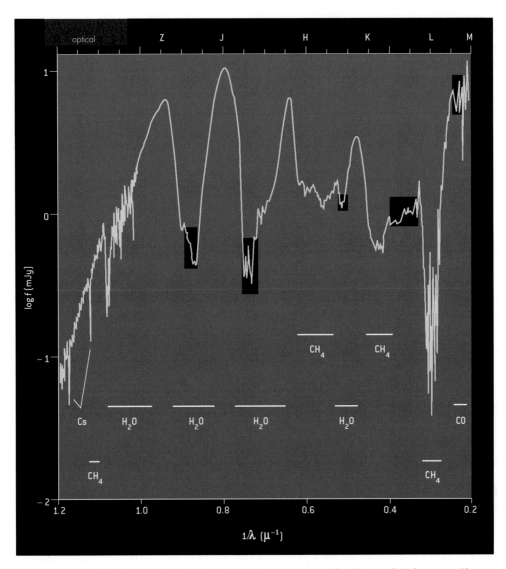

Spectrum of the brown dwarf Gliese 229 B, as measured by the Keck Telescope. The black regions correspond to spectral domains inaccessible from the ground. (After B.R. Oppenheimer *et al.*, *Ap. J.*, **502**, 832 (1998).)

The spectrum of an exoplanet, like that of any solar system object, possesses two distinct components. The first, in the ultraviolet, the visible and the near infrared, is the light from the star, reflected by the exoplanet and partly absorbed by its atmosphere. The second component is the exoplanet's own emissions. This depends on its temperature; and the lower that temperature, the further towards the far infrared the maximum flux will be. So, for Uranus and Neptune, with effective temperatures of around 50 K, the maximum flux will show at about 80 μm. Compare this with the Pegasid planets, close to their stars, at temperatures of around 1,200 K. Most of their thermal emissions are situated in the near-infrared at around 2 μm.

The atmospheric molecules detectable in the infrared are principally water, carbon dioxide, carbon monoxide, methane and ammonia. In the visible domain we also detect certain atoms – sodium, potassium, lithium and so on. The nature of the atmospheric constituents depends on the distance of the exoplanet from its star. We have seen methane and ammonia predominate at distances around a few AU, while carbon monoxide becomes important compared with CH_4 at distances closer than 0.1 AU. Nitrogen becomes more abundant than NH_3 within a distance of 0.05 AU. Various kinds of molecule can also condense out in the atmosphere of an exoplanet, considerably modifying its spectrum. This may be the case with water, silicates, titanium oxides, or even iron in solid form. To date, only a few spectra of brown dwarfs or Pegasids have been obtained in the visible and near-infrared domains; and most of these carry the signatures of water and methane. Thorough modelling of spectra is certainly important if we are to understand the nature of atmospheric opacity, which directly influences the thermal structure of exoplanetary atmospheres. Future studies of the spectra of these planets may well reveal the signatures of minor constituents.

5.12 TOWARDS NEW TYPES OF EXOPLANET

Ocean planets? Super-Earths?

There may exist other types of planet as well as those terrestrials and giants familiar to us. Concepts in the minds of theoreticians include 'ocean planets', and, deriving from recent observations yet still little understood, 'super-Earths'. The latter were detected, orbiting close to their stars, in autumn 2004.

In 2003, French astronomer Alain Léger advanced a new theoretical concept of exoplanets. He imagined a giant planet rather like Uranus, with a nucleus consisting mainly of ice and with some rocky material, starting to migrate inwards towards its star. As we have seen, this process seems to have drawn many giant exoplanets quite close to their stars. What would happen to Uranus as the ambient temperature begins to climb? Computer models show that the planet might be covered with an ocean of liquid water at temperatures between 0° and 100° C. If such exoplanets exist, exobiologists considering the possibilities of life-forms elsewhere will be extremely interested in them. All we have to do now is find these elusive objects!

With no ocean worlds in view, another kind of planet hit the headlines in autumn 2004. Three 'small' exoplanets, with masses between 14 and 20 times that of Earth, were discovered orbiting close to their stars. One of them – detected by Michel Mayor's team – is at least 14 times as massive as Earth, and orbits the star μ Arae, 50 parsecs distant. This planet is 0.1 AU from its star. The

The Earth – here imaged by Meteosat – is the most massive of the terrestrial planets orbiting the Sun. However, there could be even more massive exoEarths elsewhere in the Universe.

other two planets – found by Geoffrey Marcy's team – are scarcely more massive, and orbit near the stars Gliese 436 in Leo, and 55 Cancri, the latter of which is already known to have a multi-planet system.

In July 2006, six exoplanets with masses less than 17 Earth masses were known. Three of these – discovered by Lovis and colleagues using velocimetry – have masses respectively 10, 11 and 18 times that of Earth, and periods of 9, 32 and 197 days. An even less massive exoplanet of 5.5 Earth masses, orbiting at about 2 AU from a cool M-type star, was discovered by Jean-Philippe Beaulieu's team, using the microlensing method. How could such planets have been

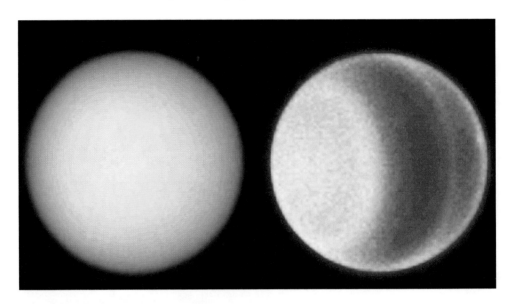

Uranus, imaged by the Hubble Space Telescope in 1997. The left-hand image was taken at 5470 Å, and is close to what the human eye would see from an approaching spacecraft. The right-hand image, at 6190 Å, shows absorption by methane molecules in Uranus's atmosphere. Here we are observing high-altitude levels. If a planet like Uranus, with its rocky and icy core, began to migrate inwards toward its star, it might eventually be covered with an ocean of liquid water as temperatures increased to the range 0°–100° C.

formed? The migration hypothesis is difficult to justify here, since planets originating in the outer part of the star's system would be much more massive. These are perhaps the 'super-Earths', assembled near their stars around a metal-silicate nucleus, within a disk of far greater mass than that which gave rise to our solar system. But this remains to be proved.

6

Life in the Universe

At present, the Earth is the only place where we know that life exists. However, we cannot deny the possibility that life may also be present on some exoEarth. Studies are targeting exoplanets which might possess liquid water – an apparently indispensable element for the emergence of life.

The Pacific Ocean. The presence of oceans has led to the emergence of life on Earth, and to the temperate climate which ensures the habitability of our planet.

6.1 HOW DO WE DEFINE LIFE?

Life on Earth: based on nucleic acids and proteins

For thousands of years our reflections on possible extraterrestrial life-forms have instinctively focused on beings not unlike those on Earth, and even on the idea of civilisations with which we might some day be able to communicate. If life has indeed begun on other worlds, including those within the solar system, there is nothing to indicate that it should in any way develop along the lines of the Earthly organisms with which we are familiar. So, before embarking upon any discussion on the subject of extraterrestrial life, we must define life.

What is living matter? Biologists tell us that it has to fulfil three criteria: self-reproduction (normally in identical form); evolution through (accidental) mutation; and self-regulation within the ambient environment (ensuring growth and the preservation of the living entity).

On Earth the cell is the structural unit of organisation in living systems. Analysis of living matter shows that all living systems employ the same types of molecule: nucleic acids and proteins. The principal nucleic acid is deoxyribonucleic acid (DNA), constituting the chromosomes and regulating the multiplication and functioning of cells. Chromosomes carry the genetic information, through DNA. The DNA double helix is built from a sequence of several thousand nucleotides, using only four bases, always associated in pairs: cytosine

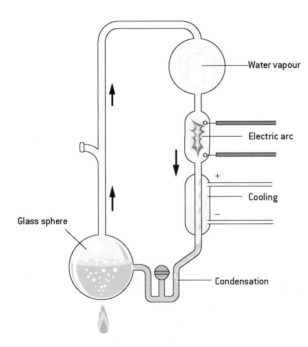

A sketch of the apparatus used by Miller and Urey to simulate prebiotic chemical evolution. This historic experiment, carried out in 1953, showed that it was possible to synthesise organic and even prebiotic molecules from a gaseous mixture of hydrogen, methane, ammonia and water vapour subjected to electrical discharges. (After P. De la Cotardière, *Dictionnaire de l'Astronomie*, Larousse, 1999, modified).

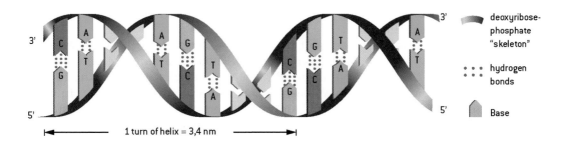

1 turn of helix = 3,4 nm

deoxyribose-phosphate "skeleton"

hydrogen bonds

Base

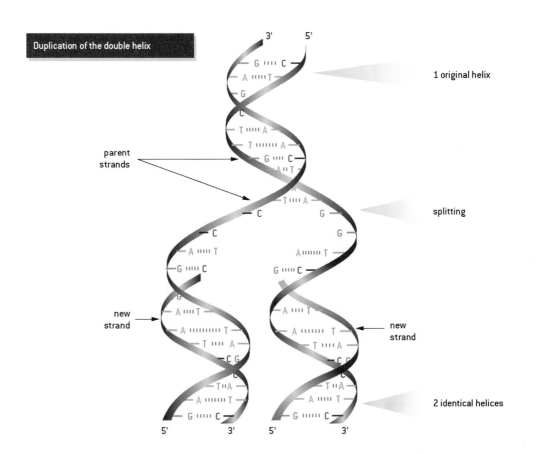

Duplication of the double helix

parent strands

new strand

1 original helix

splitting

new strand

2 identical helices

The DNA molecule and its splitting into two identical daughter molecules.

(C) with guanine (G), and thymine (T) with adenine (A). By 'unzipping' its two strands, DNA can create two identical helices, each one built upon one of the strands of the 'mother' molecule: the process of self-reproduction.

Proteins are composed of twenty different amino acids, all of which have been detected in certain meteorites. These 'prebiotic' molecules – so called because they are precursors of life without themselves being living matter – exist elsewhere in the Universe.

In 1953 the famous experiment by Stanley Miller and Harold Urey showed that laboratory synthesis of amino acids is possible in a reducing atmosphere (including H_2, CH_4 and NH_3) subjected to electrical discharges. Since the 1970s, complex organic molecules have been discovered in increasing numbers in very diverse cosmic environments, including the interstellar medium, circumstellar envelopes, and planetary and cometary atmospheres. A complex prebiotic chemistry certainly exists; but we now need to understand how life appeared on Earth.

6.2 WHY LIFE ON EARTH?

From prebiotic molecules to the cell: a missing link

How did life appear on Earth? Between the prebiotic molecules of amino acids – delivered perhaps by meteorites or synthesised in the oceans – and the most ancient fossils dating back to the earliest times, there is a gap in our knowledge which we cannot at present fill. In an attempt to supply this missing link, many laboratory experiments have been undertaken, investigating the polymerisation

Hydrothermal vents in the ocean depths.
These may well favour the synthesis of complex molecules.

M (M_E)	Venus 0.9	Earth 1	Mars 0.1
internal energy	strong	strong	weak
impact rate	high	high	weak
primitive atmosphere	dense (> 100 b)	dense (> 100 b)	tenuous (< 1 bar)
early water	vapour	liquid	mostly solid (+ liquid phase?)
present water	vapour	liquid	(liquid phase ?) permafrost
early CO_2	gas, very abundant	very abundant	fairly abundant
present CO_2	gas, very abundant	very abundant	weak gas + solid
greenhouse effect	runaway (T > 300 K)	moderate (T > 300 K)	very weak (T > 4 K)

Characteristics of terrestrial planets compared.

Stromatolites in western Australia.

of prebiotic molecules in an aqueous environment (see section 6.1). Like its neighbours Venus and Mars, the Earth must have had a primordial atmosphere rich in carbon dioxide, nitrogen and water. This somewhat reducing atmosphere might seem *a priori* unfavourable to the synthesis of prebiotic molecules, despite the presence of liquid water. However, certain other conditions, such as the existence of hydrothermal vents deep within the oceans, could have promoted the creation of complex molecules. Another possible source involves the delivery of such molecules by impacting meteorites and comets – a process which certainly contributed to the development of life on Earth and, as a result, the establishment of the primitive terrestrial atmosphere. We know that amino acids are present in certain meteorites, and that comets are rich in complex organic molecules. The Rosetta space mission, now on its way to an encounter with comet Churyumov–Gerasimenko, will be able to tell us in a few years' time if this comet also contains amino acids.

Of one thing we are sure: life on Earth began in its oceans. One piece of evidence for this can be seen in the stromatolites of western Australia. Stromatolites are stratified calcareous remnants of the physico-chemical activity of the first bacteria, dating from more than 3.5 billion years ago. Only after this era, with the advent of photosynthesis – which allowed the bacteria to transform carbon dioxide into oxygen and sugar – did Earth's atmosphere become progressively richer in oxygen. This was partly dissociated by solar ultraviolet radiation to form the famous stratospheric ozone layer, which now serves to protect us from that same harmful radiation. Life was therefore able to move onto the land and develop there, approximately 600 million years ago.

Life on Earth: chance, or inevitability? It is still an open question. Nothing suggests that life *had* to be; but we cannot ignore the tendency for the simplest molecules, in various astrophysical environments, to participate in a complex prebiotic chemistry.

6.3 SEARCHING FOR LIFE ON MARS

A myth that lasted a century

In 1877 the Italian astronomer Giovanni Schiaparelli announced a discovery which caused a sensation: the observation, on the surface of Mars, of a network

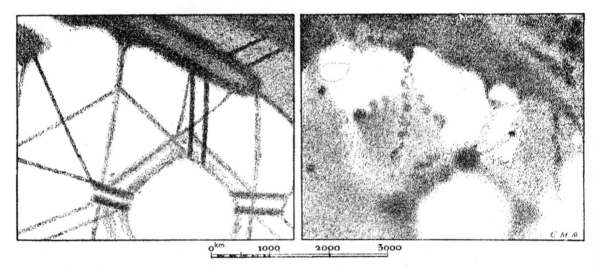

The Elysium Planitia region of Mars: a comparison of observations made by Giovanni Schiaparelli between 1877 and 1890, and by E.M. Antoniadi between 1909 and 1926. It can be seen that, with better spatial resolution (right), structures thought to be linear become irregular, disproving the theory that they are martian canals. (After E.M. Antoniadi, *The Planet Mars*, Burillier, 1930.)

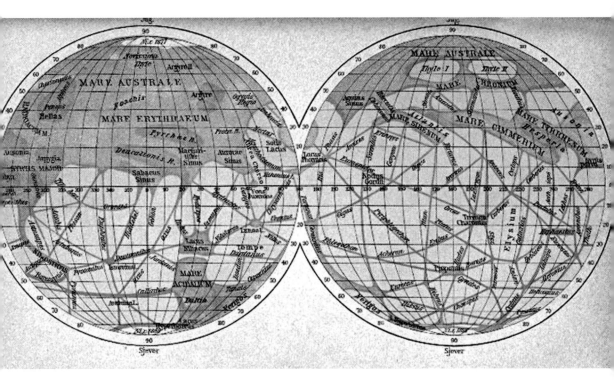

A chart of the martian surface drawn by Schiaparelli in 1879. The rectilinear structures correspond to the 'canals' interpreted by some as signs of intelligent life.

of linear structures which he called *canali*. Although Schiaparelli was careful not to draw conclusions from the observations, others did. In the USA, Percival Lowell assumed that these were 'canals': not random features, but the work of an intelligent, drought-stricken civilisation, channelling water for irrigation. Lowell devoted his life to the study of Mars, from the observatory he had built for the purpose, and which bears his name. In spite of reservations and opposition from other astronomers such as Eugène Antoniadi, the myth of intelligent life on Mars persisted well into the first half of the twentieth century. Only in the 1960s was it finally dispelled, by the first images from the spacecraft Mariner 4, in orbit around Mars. They showed that the 'linear' structures were merely an optical illusion.

Nevertheless, the question remained: could life exist on Mars, now or in the past, perhaps in the form of microorganisms? One of the main aims of the ambitious Viking project, consisting of two probes launched by NASA in 1975, was to seek an answer to this question. The Viking orbiters, carrying identical instrumentation, orbited Mars for more than two years, while the Viking landers both functioned on the surface for several years. This mission was not only a technological achievement, but also provided data about the martian atmosphere and surface – information that is still consulted.

Among the Viking landers' onboard experiments was one which captured the public imagination. Its object was to search for signs of biological activity, using various techniques. Gas emitted from soil samples in a nutritive medium was observed and analysed, and experiments tested for any photosynthetic activity from carbon-based compounds. Results were ambiguous at first, but researchers eventually came to the conclusion that biological activity was absent at the sites where the Viking landers stood. They attributed the lack of any organic molecules at the surface to solar irradiation, which would undoubtedly have destroyed them.

So, the surface of Mars, exposed to the Sun's rays, is not the place where we should search for traces of fossil or existing life. However, this does not rule out their presence below the surface of the Red Planet, or in locations on Mars which solar radiation never reaches.

6.4 THE EARTH: A HABITABLE WORLD

It all began with carbon and liquid water

With the help of what we have already learned – or, sometimes, supposed – about the emergence of life on Earth, we can now try to determine those factors the absence of which would have ruled out the possibility of it ever arising. First and foremost, the essential elements are carbon and liquid water.

Why carbon? Our observations of interstellar molecules teach us that it is the carbon atom which is the basis of the most complex molecules. Why should this be so? The answer lies in the ability of carbon to form very strong C–C bonds, and to produce, through multiple possibilities of links with other atoms, a great variety of chemical compounds. Carbon chains are therefore capable of carrying far more 'information' than molecules formed from other elements. It is also interesting to consider silicon, just below carbon on Mendeleev's periodic table of the elements. Silicon also has the potential to forge chemical links similar to those made by carbon. It can form chains, and planar and annular structures; but it is ten times less common in the Universe than carbon, because, as a heavier element, it is the product of a later process in stellar nucleosynthesis. This is certainly the reason why only a few silicon-based diatomic molecules have been found. For the purposes of complex chemistry, carbon seems to be the undisputed basis.

Why did liquid water play such an important part in the emergence and development of life? It has a specific and essential property: due to its high dipole moment it is able to electrically dissociate molecules into positive and negative ions. So, water is an excellent solvent – which was of great importance in the early history of the Earth. Water could dissolve carbon dioxide (CO_2), which reacted with calcium oxide (CaO) to form calcium carbonate ($CaCO_3$ – calcite, or chalk). Liquid water is therefore the most favourable *milieu* for chemical

$$CH_3 - (CH_2)_n - COOH$$

carboxylic acid

$$HOOC - (CH_2)_n - COOH$$

dicarboxylic acid

$$\begin{array}{c} OH \\ | \\ R - CH - COOH \end{array}$$

hydrocarboxylic acid

naphthalene

phenanthrene

anthracene

acenaphthene

fluoranthene

pyrene

indane

thiophene

$$\begin{array}{c} R \\ | \\ NH_2 - CH - COOH \end{array}$$

aminoalkanoic
acid

$$\begin{array}{c} R \\ | \\ R - NH - CH - COOH \end{array}$$

N-alkylamino-
alkanoic acid

$$\begin{array}{c} H_2 \\ C \\ H_2C \quad CH_2 \\ \backslash \quad / \\ NH - CH - COOH \end{array}$$

amino
cycloalkanoic acid

$$\begin{array}{c} COOH \\ | \\ (CH_2)_n \\ | \\ NH_2 - CH - COOH \end{array}$$

amino
dialkanoic acid

adenine

guanine

xanthine

hypoxanthine

uracile

Some organic compounds found in the Murchison meteorite. Analysis has revealed more than 500 organic compounds and 70 amino acids in this meteorite. (After M.E. Lipschutz and L. Schultz, 'Meteorites', in *Encyclopaedia of the Solar System*, Academic Press, 1999).

reactions and biochemical processes, and for the transporting of very complex microorganisms. Could there be alternatives to the aqueous environment in the Universe? Some scientists have studied the properties of other solutions such as water–ammonia or water–hydrocarbon mixtures, but these seem incompatible with the organic reactions which have led to the existence of life as we know it. The emergence of life in liquid water is therefore not due to mere chance. The physical and chemical properties of water make it the most propitious environment in which life can develop.

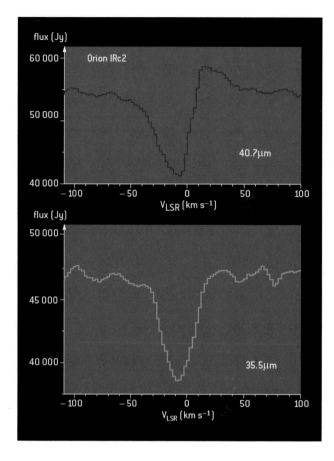

The detection of two water vapour lines in front of the molecular cloud Orion Irc-2 by the SWS infrared spectroscope on the ISO satellite. (After E. Van Dishoek et al., ESA SP-427, 1999.)

It is worth mentioning that the H_2O molecule – composed of two atoms which are very abundant in the Universe – has also been detected (in its alternative forms of ice and vapour) in all types of places: from the Earth's neighbouring planets, through stars both young and old, to distant galaxies – and even in sunspots!

6.5 LIQUID WATER IN THE PAST HISTORY OF MARS?

Numerous indications that water flowed on Mars

As we have stated, liquid water is an important element in the development of life. In the laboratory, the synthesis of prebiotic molecules is possible in the presence of liquid water, given a source of energy; while on Earth, water allowed life to emerge. Current temperatures and pressures on the surface of Mars are such that liquid water cannot persist there, and it therefore exists in the form of vapour or ice. However, this was not always so: Viking images in the late 1970s caused astronomers to suspect the presence, in Mars' distant past, of liquid water flowing across its surface. There are many indications pointing to primeval water on Mars: ancient branching valleys resembling dried-up networks of water-courses on Earth; outflow features amid chaotic terrain, evidence of past

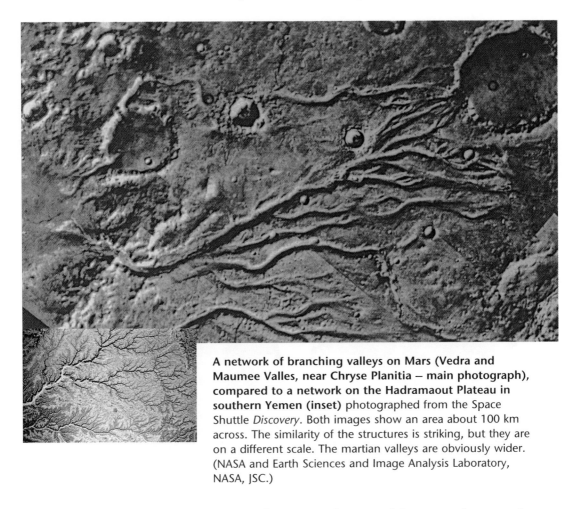

A network of branching valleys on Mars (Vedra and Maumee Valles, near Chryse Planitia – main photograph), compared to a network on the Hadramaout Plateau in southern Yemen (inset) photographed from the Space Shuttle *Discovery*. Both images show an area about 100 km across. The similarity of the structures is striking, but they are on a different scale. The martian valleys are obviously wider. (NASA and Earth Sciences and Image Analysis Laboratory, NASA, JSC.)

flooding; and lobed ejecta around craters, where mud has surged outwards, suggesting the presence of a viscous or liquid subsurface element.

Later missions served to confirm the astronomers' suspicions. Mars Global Surveyor, launched in 1996, established, from precise altimetric measurements, that between ancient terrains to the south and lower-lying plains to the north there is a ridge that is constant in altitude over more than 1,000 km – possibly a former ocean shoreline. Could this ocean have covered the whole of Mars' northern hemisphere? This seems improbable. Mars Express did not detect any sediments in the northern plains, although it detected sediments on the edges of ancient cratered regions in the southern hemisphere. These deposits may well be evidence of abundant water flowing during Mars' distant past – possibly at some time in the first billion years of its history. Also, the Mars Express orbiter and the robot surface vehicles Spirit and Opportunity have revealed the existence in various locations of sulphate material, the formation of which seems to require

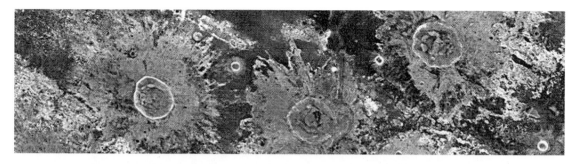

Lobed-ejecta craters in the Isidis Planitia region of Mars, photographed by Mars Odyssey in December 2002. On the plains of Mars' northern hemisphere, ejecta around this type of crater are particularly extensive, suggesting the presence of large amounts of water ice at the moment of impact.

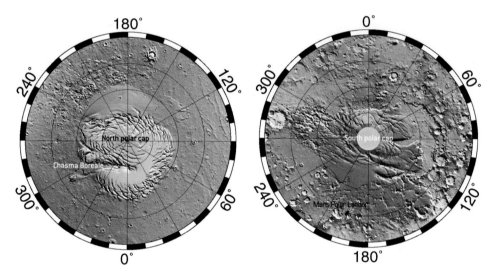

A topographical chart of Mars. The altitude of the line separating the northern and southern regions appears to be constant over distances of the order of 1,000 km. The southern hemisphere (right), which is more heavily cratered, has an appreciably higher mean altitude than the northern plains.

the presence in past times of large quantities of liquid salt water. These sulphates, having formed later, could be the vestiges of violent, episodic flows, caused perhaps by volcanic activity. How long did the water remain on the surface of Mars? Was it for long enough to permit the emergence of life? If life did evolve, then is it likely that we could find traces of it, whether or not it still exists? The exploration of Mars is entering a new phase, and there is no doubt that it will constitute one of the great scientific challenges of the twenty-first century.

6.6 EUROPA AND TITAN: HARBOURING LIFE IN THE OUTER SOLAR SYSTEM?

Is there an ocean below the surface of Europa? Is Titan's atmosphere conducive to the emergence of life?

It seems hardly probable that gas planets like Jupiter and Saturn, or ice giants like Uranus and Neptune, could offer 'life-friendly' environments. If they possess any solid surface to their internal cores, pressures there will be so enormous that any kind of life as we can imagine it could not survive. It is possible to envisage some kind of prebiotic, hydrocarbon-based chemistry occurring in the clouds of methane, ammonia and water vapour surrounding the giant planets, at levels where pressures range from 1 bar to a few bars. However, any macromolecules resulting from such processes would, in the absence of a solid surface, tumble downwards as gravity drew them towards inevitable destruction in regions at higher temperatures and pressures.

We must therefore concentrate on bodies with solid surfaces: the satellites of these planets. Two of them merit particular attention: Europa, a Galilean satellite of Jupiter, and Titan, Saturn's largest moon. Titan, like the Earth, has a dense atmosphere. Europa, in common with all of Jupiter's satellites, has no stable atmosphere. Infrared spectroscopy has shown that its surface is covered with water ice: the first surprising close-up images of Europa's surface, transmitted by Voyager 1 in 1979, showed a complex network of intersecting linear structures. Fifteen years later, the Jupiter orbiter Galileo sent more images. The 'floes' delineated by these markings on Europa seem to have moved relative to each other, upon some viscous or even liquid medium beneath. The most likely candidate is water – probably liquid, salt water. This subsurface ocean may

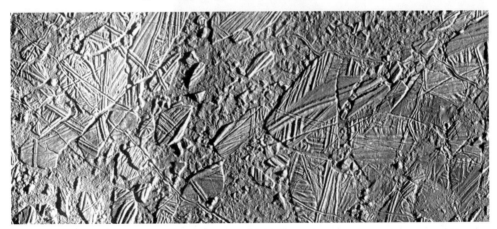

The surface of Europa, imaged by the Galileo spacecraft. The interlaced linear structures are the edges of 'floes' which have probably been displaced with respect to each other on top of a viscous or even liquid *milieu*: a sub-surface ocean.

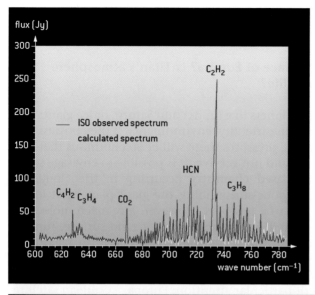

The infrared spectrum of Titan, from ISO/SWS (1997), in the range 600–780 cm^{-1} (12.82–16.6 μm). (After A. Coustenis *et al.*, ESA SP-419, 255, 1997.)

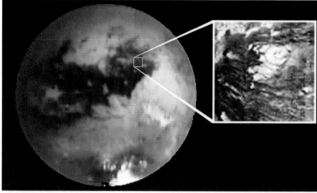

The surface of Titan, imaged by the Cassini orbiter. The false colours represent characteristics of the atmosphere (red) and surface (blue and green) invisible to the human eye.

also be responsible for Europa's magnetic field, discovered by the Galileo spacecraft. This field probably results from an interior 'dynamo' effect. Where there is water, perhaps, there may be an environment favourable to life. There is of course the problem of the depth of Europa's icy crust: current models suggest it is at least 10 km thick, and maybe much more. The great challenge will be drilling into it, and scientists are contemplating a preparatory probe, which will orbit Europa, taking accurate altimetric readings and studying its gravitational field.

Titan, in the far-off environment of the saturnian system, has intrigued astronomers since the early twentieth century, when observations by Comas Sola disclosed the existence of a dense atmosphere. Titan is the only satellite in the solar system that possesses one. Telescopic investigations, and later studies by Voyager 1 in 1981, showed that the surface of Titan is permanently veiled beneath a dense layer of cloud, and that its atmosphere is composed mainly of molecular nitrogen. Methane was also discovered, as well as large numbers of hydrocarbon molecules and nitriles – products of the dissociation of CH_4 and N_2. Here are building blocks which may contribute to the formation of more complex prebiotic molecules. Could Titan's atmosphere be a hospitable site for the emergence of life? One considerable difference between this atmosphere and that of the Earth, however, is that on Titan the temperature is much lower, so that any chemical reactions occurring there would proceed far more slowly. Nevertheless, Titan remains a fascinating object, and a prime objective of the ambitious Cassini–Huygens mission was its exploration. The Huygens lander touched down successfully on 14 January 2005, and returned the first images of Titan's surface, revealing probable traces of ancient lakes of liquid hydrocarbons, now dried out. The Cassini orbiter continues its years-long exploration of Saturn and its system.

6.7 THE FIRST BUILDING BRICKS: INTERSTELLAR CHEMISTRY

More than a hundred molecules have been identified in space

Carbon chemistry is, as has already been stated, a very common feature of the Universe. Many complex carbon-based molecules have been detected by radio astronomy or millimetric methods, as their size dictates that their vibrational modes are at low frequencies. Most of these polyatomic molecules have been found in the interstellar medium. What are its constituents? First, there are dense molecular clouds at very low temperatures (10–20 K), and these are favoured sites for the formation of interstellar molecules. At such low temperatures, any gas molecule encountering a solid grain will adhere immediately, and an icy mantle will be formed. If organic molecules are present in this mantle, they will be subjected to ultraviolet radiation by nearby stars, accelerating their evolution into even more complex refractory molecules. The process continues if the grain is part of the denser and warmer circumstellar environment of a nascent star. Complex molecules formed at low temperatures come together in the gaseous phase, enriching the ambient protoplanetary medium. Similarly, the circumstellar envelopes of late-stage stars can contribute complex organic molecules to the surrounding interstellar medium.

More than 120 interstellar molecules are now known. The largest of them, comprising thirteen atoms, is $HC_{11}N$. The most voluminous polyatomic molecules all consist of carbon-based chains, illustrating the capacity of carbon to produce a complex chemistry. As well as these individual molecules, radio astronomers have also detected, over the last twenty years, polycyclic aromatic hydrocarbons (PAH). These are particularly stable molecules, identifiable by their infrared spectra. They contain many atoms, arranged in 'benzene rings', and occur in abundance in star-forming regions. They provide yet more evidence of the pre-eminence of carbon chemistry in the Universe.

Hydrogenated compounds

H_2	$\mathbf{H_3^+}$

Carbon chains and rings

CH	$\mathbf{CH^+}$	C_2	CH_2	CCH	C_3
CH_3	C_2H_2	$l\text{-}C_3H$	$c\text{-}C_3H$	CH_4	$C_4?$
$c\text{-}C_3H_2$	$l\text{-}C_3H_2$	C_4H	C_5	C_2H_4	C_5H
$l\text{-}H_2C_4$	HC_4H	CH_3CCH	C_6H	C_6H_2	HC_6H
C_7H	CH_3C_4H	C_8H	C_6H_6		

Compounds containing hydrogen, oxygen and carbon

OH	CO	CO^+	H_2O	HCO
HCO^+	$\mathbf{HOC^+}$	C_2O	CO_2	$\mathbf{H_3O^+}$
$\mathbf{HOCO^+}$	H_2CO	C_3O	$HCOOH$	CH_2CO
$\mathbf{H_2COH^+}$	CH_3OH	CH_2CHO	HC_2CHO	C_5O
CH_3CHO	$c\text{-}C_2H_4O$	CH_3OCHO	CH_2OHCHO	CH_3COOH
CH_2CHOH	$(CH_3)_2O$	CH_2CHCHO	CH_3CH_2CHO	CH_3CH_3OH
$(CH_3)_2CO$	$HOCH_2CH_2OH$	$C_2H_5OCH_3$		

Compounds containing hydrogen, nitrogen and carbon

NH	CN	NH_2	HCN	HNC	$\mathbf{N_2H^+}$
NH_3	$\mathbf{HCNH^+}$	H_2CN	$HCCN$	C_3N	CH_2CN
CH_2NH	HC_3N	HC_2NC	NH_2CN	C_3NH	CH_3CN
CH_3NC	$\mathbf{HC_3NH^+}$	C_5N	CH_3NH_2	C_2H_3CN	HC_5N
CH_3C_3N	C_2H_5CN	HC_7N	$CH_3C_5N?$	HC_9N	$HC_{11}N$

Compounds containing hydrogen, oxygen, nitrogen and carbon

NO	HNO	N_2O	$HNCO$	NH_2CHO	NH_2CH_2COOH

Sulphur and silicon compounds and other types

SH	CS	SO	$\mathbf{SO^+}$	NS	SiH	SiC
SiN	SiO	SiS	HCl	$NaCl$	$AlCl$	KCl
HF	AlF	CP	PN	H_2S	C_2S	SO_2
OCS	$\mathbf{HCS^+}$	$c\text{-}SiC_2$	$SiCN$	$NaCN$	$MgCN$	$MgNC$
H_2CS	$HNCS$	C_3S	$c\text{-}SiC_3$	SiH_4	SiC_4	CH_3SH
C_5S	FeO	$AlNC$				

Deuterium types

HD	$\mathbf{H_2D^+}$	$\mathbf{D_2H^+}$	HDO	CCD	DCN	DNC
$\mathbf{DCO^+}$	$\mathbf{N_2D^+}$	HDS	D_2S	NH_2D	ND_2H	ND_3
$HDCO$	D_2CO	$HDCS$	CH_2DOH	CD_2HOH	CD_3OH	CH_3OD
DC_3N	DC_5N	C_4D	CH_2DCCH	CH_2DCN	$c\text{-}C_3HD$	

Molecules detected in the interstellar medium and in circumstellar envelopes.
Italicised entries (such as H_2) indicate detection in UV or the visible. Bold entries (such as H_3^+) are molecular ions. Red entries have been detected only in circumstellar envelopes and (proto)planetary nebulae. Green entries have been detected in both the interstellar medium and in stars. Underlined entries are molecules detected in external galaxies. Molecules with minor isotopes have also been detected: ^{13}C, ^{18}O, ^{17}O, ^{15}N, ^{29}Si, ^{34}S, ^{33}S, and so on. (Table produced by the national Physique-Chimie du Milieu Interstellaire programme.)

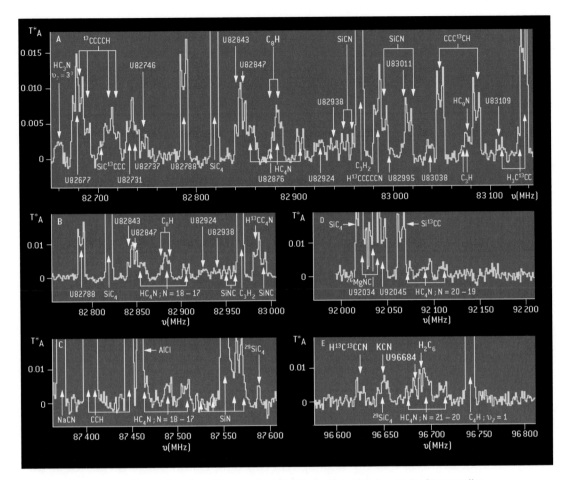

Complex interstellar molecules. Observations (in the millimetric domain) of interstellar clouds or, in these cases, stellar envelopes, reveal the presence of many complex molecules. The letter U refers to unidentified molecules. (After Cernicharo, *Ap. J.*, **615**, L145 (2004).)

6.8 THE HABITABILITY ZONE IN PLANETARY SYSTEMS

Where do we look for life? Where there is liquid water

Let us now look outwards from the solar system, towards the stars. Where might we hope to find life? We require carbon and liquid water. The first condition is easy to fulfil, since carbon chemistry is ubiquitous in the Universe. The second, though, is more restrictive. Water may well exist in other stellar systems, but it occurs exceptionally rarely in liquid form. Within the solar system, only the

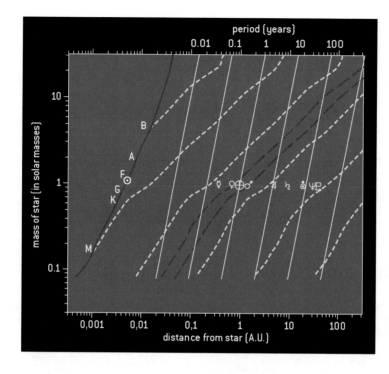

The habitability zone as a function of the mass of the star. (After Eddington, *Mission Definition Document*, ESA.)

Earth and Mars have, or have had, conditions of temperature and pressure favourable to the presence of liquid water. The phase diagram of the water molecule imposes these narrow conditions, and, at the pressures to which we are accustomed, water will be liquid at temperatures between 0° C and 100° C (273–373 K). In the case of the solar system this temperature range is reached at the heliocentric distance of the Earth (1 AU).

Returning to the solar system, within our galaxy, we ask: what are the most favourable environments for the emergence of life? Exoplanets of terrestrial type – exoEarths – could possess water in liquid form. They would have to have temperatures between 0° C and 100° C, and therefore be at a certain distance from their central stars. If these stars are of solar type, the distance in question will be, as in the case of the Earth, of the order of 1 AU. If the central star is smaller and less luminous, the distance will be reduced; while with a more massive star it is increased. The region within which water can occur in liquid form on an exoplanet is called the habitability zone (see p. 108).

Remember that this definition of the habitability zone is perhaps restrictive; it is based upon the notion of life as we understand it, and refers to the development of life on Earth. We have seen that there are other more exotic places within the solar system which might qualify, such as the subsurface ocean of Europa or the prebiotic atmosphere of Titan (see section 6.6). We have not taken these into account in our definition. We are dealing here with a

Earthrise over the Moon. The presence of a massive satellite orbiting a planet can play an important role in the development of life on that planet by stabilising its rotational axis. This has undoubtedly been a feature of the Earth–Moon system.

convenient notion to enable us to select those exoplanets which, of all those discovered, might be likely places for life to begin.

It must be noted, too, that the existence of an exoEarth within its habitability zone does not necessarily imply that life can develop there. If we consider planet Earth there are other determining factors, apart from its distance from the Sun, which have come into play to encourage the emergence and maintenance of living things. A very important factor is the presence of Jupiter. Although it is a great distance away, this enormous planet, with its powerful gravitational field, has acted as a veritable shield to protect the inner solar system from a multitude of impacts which might otherwise have compromised the first stirrings of life. Similarly, computer models tell us that the existence of our large satellite, the Moon, has served to stabilise the Earth's axis of rotation in relation to the plane of the ecliptic. If this axis had undergone a series of periodic oscillations, like those which certainly occurred in the case of Mars, the development of life on Earth would have been severely affected.

6.9 THE SEARCH FOR EXOEARTHS

Tomorrow: another Earth discovered?

We have mentioned exoEarths – but are they really there? At present they reside within the speculations of astronomers who, having discovered more than 200

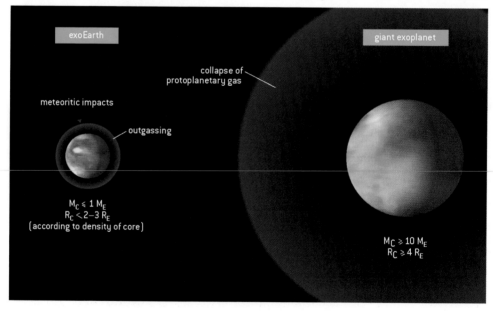

An exoEarth compared with a giant exoplanet.

The French/European COROT mission will enable space-borne instruments to detect and examine the flux from systems with exoEarths undetectable from the ground.

giant planets, think – justifiably – that smaller exoplanets, like the 'terrestrials' of the solar system, ought to exist. Exoplanets of smaller and smaller masses are being found.

Current observational techniques cannot reveal such objects to us. The current limit of detectability, using the velocimetric method by which most discoveries have been made, is approximately 15 Earth masses. This corresponds to planets the size of Uranus and Neptune. So we are still in the realms of giant planets – 'giant' here referring to planets with nuclei of 10–15 Earth masses, enabling the capture of surrounding protosolar gases. Below this limit, a planet will consist essentially of a solid core, and any atmosphere will constitute only a tiny fraction of the total mass. The maximum radius for a terrestrial planet is about two or three times that of the Earth, assuming its nucleus to be composed of ice or rock.

How can we detect exoEarths? Since their low masses preclude detection by the velocimetric method, it will be necessary to resort to the transit or microlensing methods (see section 2.8). Seen from outside the solar system, the passage of Jupiter across the Sun would involve a reduction in solar brightness of the order of 1%. In the case of the Earth the reduction would be 0.01%. When we observe other planetary systems we can measure to an accuracy of 1%, so we can detect the passage of a giant exoplanet across the face of its star. However, measuring variations in starlight a hundred times fainter is beyond our capabilities, so exoEarths cannot be detected from ground-based locations. To obtain such results it is necessary to observe from space. This is the objective of the French/European COROT mission, launched in December 2006, carrying a battery of highly sensitive CCD cameras, to be aimed at various star-fields throughout a period of several months. Astronomers hope that its continuous observation of thousands of stars may lead to the first discovery of Earth-like exoplanets.

6.10 DETECTING LIFE ON EARTH

Two clues: chlorophyll at the surface and ozone in the stratosphere

If we were observers in distant space, outside the solar system, would we be able to detect life on Earth? This is an important question, the answer to which helps us define the criteria for recognising life on any exoEarths we might find. Twice – in December 1990 and December 1992 – the Galileo spacecraft, which had been launched in 1989 on a long and tortuous journey to Jupiter, passed near the Earth. The aim was to use the Earth's gravitational interaction with the probe to accumulate enough kinetic energy to reach its distant destination. On both occasions, instruments on Galileo were pointed towards the Earth, looking for any tangible signs of life. Several positive responses were reported from the study of the planet's electromagnetic spectrum – its radiation as a function of wavelength, from ultraviolet through infrared and into the radio spectrum.

First, Galileo's infrared spectrometer studied Earth's spectrum in the near infrared. The signatures of water and carbon dioxide were much in evidence, but the instrument also detected the unambiguous presence of ozone (O_3) and methane (CH_4). It is very difficult to explain the presence of these two molecules in abundance without invoking the action of biogenic mechanisms. It is ozone, the product of photodissociation of molecular oxygen, which is the tracer, rather than oxygen itself, which is more difficult to detect spectro-scopically. So, the detection of ozone provides evidence of the presence of

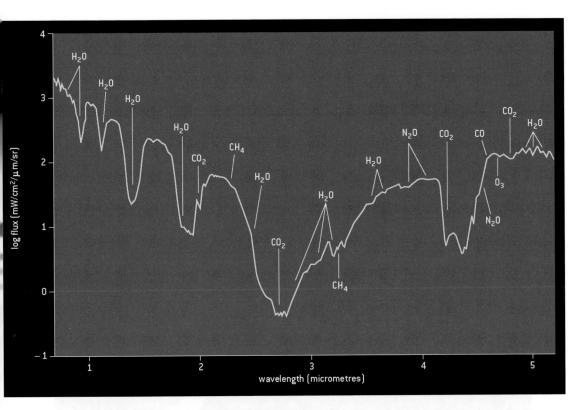

The infrared spectrum of Earth's environs, dominated by the signatures of water and carbon dioxide. (After P. Drossart *et al.*, *Plan. Space Sci.*, **41**, 551 (1993).)

molecular oxygen. Methane can persist in the Earth's atmosphere only by being constantly renewed, as it reacts very rapidly with atmospheric oxygen to produce carbon dioxide. We know of no abiotic mechanism capable of producing large quantities of methane, though this is not to say that such mechanisms cannot exist.

Next, Galileo's measurements of the spectrum of reflected sunlight over the land, at around 0.8 μm, revealed a very important characteristic: the signature of chlorophyll. This second indicator, however, might not figure as prominently in the case of exoEarths, as there is no proof that any life-forms on such planets would employ the kind of photosynthetic reactions with which we are familiar.

Lastly, Galileo's radio receiver detected strong emissions, confined to very narrow bands and very specifically modulated in frequency. These were, of course, Earth's radio and television signals. This third indicator is certain evidence of a technologically advanced civilisation.

How can we apply these criteria to the search for extraterrestrial life? Our principal tool is spectroscopy, which will allow us to analyse the chemical composition of the atmospheres, and even the surfaces, of exoplanets.

En route to Jupiter, the Galileo spacecraft made two passes by the Earth. Measuring the reflected solar spectrum over land masses, it revealed the signature of chlorophyll on our planet at around 0.8 μm.

6.11 DETECTING LIFE ON EXOPLANETS

The infrared: hunting ground for traces of life

Of what does the spectrum of an exoplanet consist? As we have seen, it has two parts. At shorter wavelengths – in the ultraviolet, the visible and the near infrared – radiation from the planet represents the reflected flux of its star. The spectral signatures of the planet's atmosphere and surface appear at their characteristic wavelengths: water, carbon dioxide and methane show strong signals in this domain. At longer wavelengths, in the mid-infrared and far-infrared, we find the planet's own emissions – signals of its temperature. An exoEarth orbiting at 1 AU

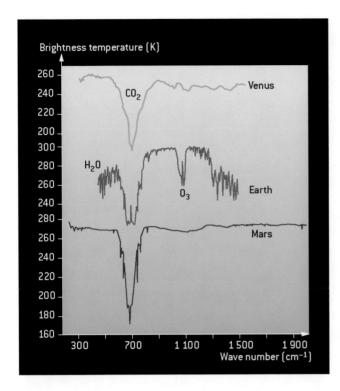

The infrared spectrum of the terrestrial planets of the solar system, between 5.2 μm (1,900 cm^{-1}) and 33 μm (300 cm^{-1}). There are numerous characteristic signatures of gases present in their atmospheres: mostly CO_2 for other terrestrial planets, and O_3 for Earth. The vertical scale (brightness temperature) shows the flux emitted by the planet.

from its (solar-type) star will have a temperature of about 270 K, and its radiation will register preferentially at approximately 10 μm.

So, if we wish to observe the spectrum of an exoEarth we should be looking for radiation in the mid-infrared at around 5 and 30 μm. What molecules can we identify within this domain? They include water (H_2O) at 6 μm and beyond 20 μm, carbon dioxide (CO_2) at 15 μm, methane (CH_4) at 7.7 μm, ammonia (NH_3) at 10 μm, and ozone (O_3), with an intriguing signature showing well at 9.7 μm.

It is of interest to compare the spectra of the three 'terrestrial' planets with atmospheres in the solar system: Venus, the Earth and Mars. The sulphuric acid clouds blanketing Venus, with an average pressure of approximately 1 bar, emit thermal radiation from the planet. All three planetary spectra show a very intense signature of CO_2 at 15 μm. The Earth's spectrum shows clear evidence of water, and the strong signal of ozone. However, the signal of methane at 7.7 μm is veiled by that of water. A distant observer might well conclude that molecular oxygen is present in large quantities. Would this necessarily indicate the presence of life? It might not be a certainty, but such a signal would suggest it as a serious probability.

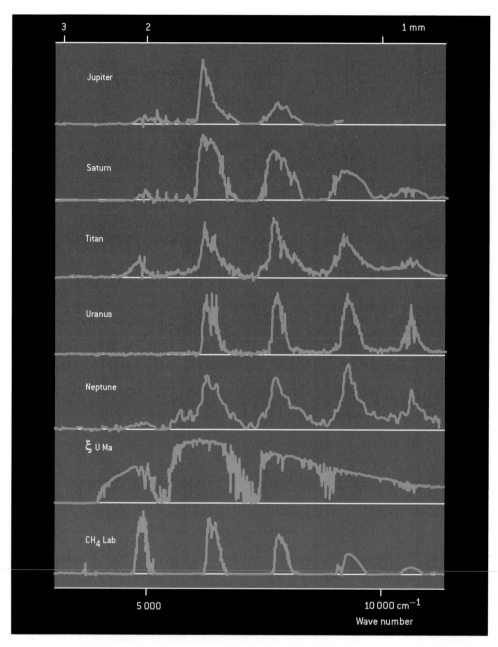

The near-infrared spectrum of the giant planets of the solar system, as recorded by high-resolution spectroscopy from Earth. The spectrum of the star ζ Ursae Majoris serves to indicate spectral domains where absorption by the Earth's atmosphere is important. It can be seen that the planetary spectra are dominated by absorption of methane. The laboratory spectrum of methane is shown at the bottom. (After H. Larson, *Annual Rev. Astron. Astrophys.*, **18**, 43 (1980).)

6.12 THE SEARCH FOR EXTRATERRESTRIAL CIVILISATIONS

SETI: a dedicated search for extraterrestrial radio signals

The continuous stream of radio emissions from planet Earth betrays the presence of an advanced technological civilisation. For decades, radio astronomers have been investigating the likelihood of communicating with possible extraterrestrial civilisations, using radio signals. The favoured frequency is that of a particular transition of neutral hydrogen at 1,420 MHz, corresponding to a wavelength of 21 cm. Why this frequency? Atomic hydrogen is the most abundant element in the Universe, and the Earth's atmosphere – and the atmosphere of any exoEarth resembling it – is transparent in this spectral domain. We can therefore study this frequency, without interruption, using large ground-based radio telescopes.

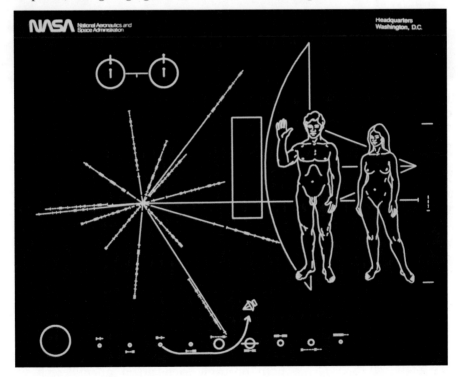

The plaque attached to the Pioneer 10 and Pioneer 11 spacecraft, bearing a message to a possible extraterrestrial. At top left is a diagram showing the motions of the proton and the electron of an atom of neutral hydrogen; at centre left is a diagram of the periods of fourteen pulsars, and their positions with respect to the solar system; along the bottom is a representation of the solar system, with the trajectory of the spacecraft indicated, showing that it originates from Earth; and to the right, a picture of the spacecraft behind a pair of human beings to the same scale.

The spherical reflector of the great radio telescope at Nançay, in the Sologne region of France. This instrument can be used to transmit or receive radio signals as part of the search for extraterrestrial civilisations.

What is the probability that life exists elsewhere in the Universe?

What is the probability that an extrasolar planet harbours some form of life? This question was being asked long before the discovery of exoplanets. Let us try to apply some numbers to this problem, in the manner of Frank Drake and Carl Sagan in the 1960s and 1970s. If N is the number of planets in our galaxy where life-forms currently exist, then:

$$N = N^* \times f_p \times n_p \times f_l$$

In this equation – a version of Drake's Equation – N^* represents the number of stars in our galaxy; fp is the percentage of stars possessing planets; n_p is the number of planets per system capable of supporting life; and f_l is the percentage of planets upon which life actually appears. Of these four parameters, N^* and f_p are both relatively well-known quantities. If we concentrate on small stars not too different from the Sun, N^* is about 10^9. If we take into account what has been discovered in the last ten years, then f_p must be at least of the order of 10%. We have absolutely no idea what the values of the other two parameters might be. Depending on how pessimistic or optimistic we are, we should find that there exist within the Milky Way between a few dozen and hundreds of billions of planets where life exists, or has existed!

To this end, in the 1960s giant radio dishes began to survey the sky at 21 cm. In 1984, project SETI (Search for Extraterrestrial Intelligence) was launched. Originally funded by NASA, it continues with private backing. This vast international undertaking finances various research projects – most of them involving listening out for 'intelligent' radio signals from elsewhere in the cosmos. Although the search has yet to bear fruit, it does indicate the considerable collective interest in this quest, which extends well beyond the scientific community. There have been other symbols of our interest in extraterrestrial beings: the messages attached to the spaceprobes Pioneer 10 and 11, and Voyager 1 and 2, sent into space in the 1970s, and all destined to leave the solar system. These 'messages in a bottle' may have little chance of ever being discovered, but what they *do* illustrate is the extent of our determination to pursue the quest.

7
Future projects

The space missions and telescopes of the future will ensure the rapid expansion of exoplanetary exploration during the next decade. From 2006 the COROT mission will be searching for exoplanets using the transit method.

An artist's impression of the Atacama Large Millimetre Array (ALMA). This instrument – at present under construction in Chile – will be able to conduct millimetric and submillimetric studies of protoplanetary disks.

7.1 OBSERVING THE FORMATION OF PLANETS IN PROTOPLANETARY DISKS

Radio and infrared observations investigate the formation of planets

How are planetesimals made from pieces of rock and ice just metres across? Do giant planets really migrate inwards through protoplanetary disks, leaving gaps in their wake? How are terrestrial planets built within habitable zones? Answers to these questions will remain elusive as long as we have practically no direct observations of what actually takes place inside a protoplanetary disk. Since these disks are opaque to visible light we must resort to the infrared and radio domains to pursue such questions. Spatial resolution and sensitivity in those domains are still not sufficiently high to obtain images of disks. In the infrared, instruments working now and in the future with arrays such as the ESO's Very Large Telescope Interferometer (VLTI) may reveal something of the activity in central regions near stars, but imaging will be very problematical. Radio astronomers are anticipating the completion, within the next decade, of the Atacama Large Millimeter Array (ALMA) – an interferometer consisting of fifty telescopes designed for millimetric and submillimetric surveys. A joint project of Europe, the USA, Canada and Japan, ALMA is currently under construction in the Atacama Desert of northern

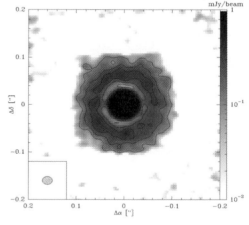

An ALMA simulation of gap clearing in a protoplanetary disk. (Computer simulation after Wolf *et al.*, *Ap. J.*, **566**, L97 (2002).)

ALMA is an interferometer comprising 50 telescopes operating in millimetric and submillimetric wavelengths. It will be in service in 2010. Now under construction by Europe, the USA, Canada and Japan in the Atacama desert in Chile, ALMA will enable a complete study of the size, dynamics, composition, etc. of protoplanetary disks. ALMA will not 'see' protoplanets directly, but will be able to image the gaps they create in their disks.

The SKA radio interferometer is scheduled for completion between 2015 and 2020. Not only will it be able to study the structure of protoplanetary disks, but also the various sizes of the dust grains in different regions of the disks. This is an artist's impression of one possible concept.

Chile, at an altitude of 5,000 metres. Uniquely, it will be able to study protoplanetary disks in all their aspects: size, dynamics, chemical composition, temperature, and the nature of their dust. Although ALMA will be unable to 'see' even giant protoplanets, it ought to be possible for it to image the gaps created in protoplanetary disks by planets orbiting (like the solar system giants) at a few AU from their stars. ALMA's results will be a valuable resource, enabling us to refine theories and simulations of the formation of giant planets. Another project – less well advanced, but of great interest for the study of planets in their disks – is the Square Kilometre Array (SKA). This is another radio interferometer, operating in the metric and centimetric domains. The SKA will provide essential data on the distribution of dust grains of various sizes, around the critical diameter of 1 cm, in different regions of disks. It will also study disk structure, investigating disk gaps and spiral waves created by protoplanets, but within the habitability zone at distances of less than 1 AU from stars. Since such protoplanets will have orbits of the order of 1 year, it may be possible, by observing at intervals of a few months, to detect changes in the structure of the disk caused by the movement of the planets. Researchers anxious to pursue direct studies of the formation of Earth-like planets will, however, have to be patient, as the ambitious SKA project will doubtless not be functional until 2015–20.

7.2 THE FUTURE FOR VELOCIMETRY

A new generation of high-resolution spectrometers

Accurate measurement of the radial velocities of nearby stars has led to the detection of the vast majority of known exoplanets, with the exception of those accompanying pulsars (see p. 20). Can this tried and trusted method offer us any new discoveries in the years to come?

Modern instruments can determine stellar velocities to within a few metres per second, which means that they can be used to study some 2,000 stars of a spectral type similar to that of the Sun, out to a distance of a few tens of parsecs.

The 3.6-metre telescope at the ESO site, La Silla, Chile. The HARPS spectrograph is installed on this instrument. HARPS' accuracy combines with the large size of the telescope to ensure important advances in the search for exoplanets using the velocimetric method.

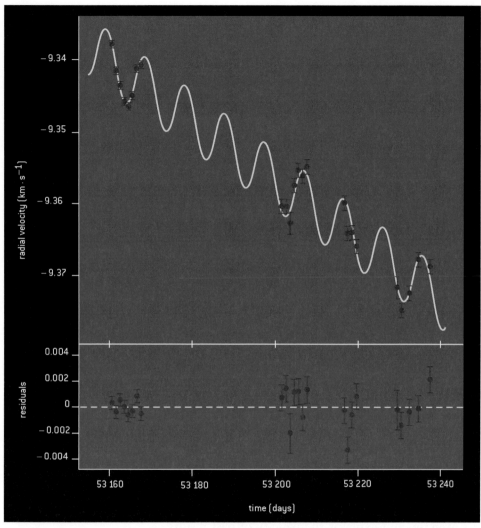

The velocimetric curve of the 'super-Earth' (17 Earth masses) detected by HARPS in August 2004. The residuals of the fit (departures of the points from the curve) are shown at the bottom.

In the not-too-distant future we will have discovered most of the giant exoplanets (those of masses at least half of Jupiter's mass) with orbital periods of less than 1 year and within 50 parsecs of the Sun. We remarked elsewhere that after a period which saw a rapid succession of discoveries, the detection rate has slowed somewhat during the last few years; massive, short-period Pegasids have been the easiest to find. Now, recently developed instruments are opening new perspectives: for example, the High Accuracy Radial Velocity Planetary Search (HARPS) spectrometer, mounted on the 3.6-metre ESO telescope at La Silla in Chile, and in use since 2004. HARPS works on the same principle as its predecessor, simultaneously measuring the Doppler shifts of a large number of spectral lines in the visible domain. It is capable of observing 1,000 stars beyond 50 parsecs to an accuracy of 1 m/s. Compare the shifts in the case of the Sun, due to interactions with Jupiter and Saturn: respectively, 12 m/s and 3 m/s. HARPS' great sensitivity has placed within our reach a whole new class of giant exoplanets: exoSaturns, or perhaps even less massive planets. A spectacular early illustration of the capabilities of HARPS has been the discovery of a planet of 17 Earth masses (the mass of Neptune) at a distance of approximately 0.1 AU from its star. Other teams, using similar techniques, have discovered two objects of about the same mass. These new-style exoplanets have certainly intrigued the theoreticians. What of their composition and their origins? Are they the first members of a new class of object? There is no doubt that velocimetry, pushed to the limits of its detective capability, holds many more surprises in store!

7.3 THE ASTROMETRY OF TOMORROW

New horizons revealed through interferometry

For a long time, astrometry has been unable to detect low-mass objects orbiting stars. It may, however, take on a new lease of life in the decade to come. As we have seen, this method involves the direct measurement of the displacement of a star due to the presence of a companion. The difficulty lies in determining very distant and supposedly 'fixed' reference stars, against which the movements of the target star can be measured to a sufficient degree of accuracy. If we take the example of the Sun, its motion due to the effects of Jupiter and Saturn, as seen from a distance of 15 parsecs, would be 0.33 and 0.18 milliarcseconds. Compare this with the Hipparcos catalogue of stellar positions – the most accurate to date – which has an accuracy of about 4 milliarcseconds. This shows the difficulty of this exercise, and explains the paucity of past results.

To improve the accuracy of astrometric measurements we need to resort to interferometry, which uses the combined signals from a number of telescopes to synthesise images at very high angular resolution. Employed in radio astronomy for many decades, very-long-baseline interferometry (VLBI) can achieve angular resolutions of the order of 0.1 milliarcseconds. However, this is restricted to very intense radio sources, and the possible range of targets is considerably reduced.

We must therefore look to the interferometers of the future, working in the visible and near-infrared. The first prototypes of these instruments are now being developed. Several of them seem very promising. The interferometer of the Large Binocular Telescope (LBT), consisting of two telescopes on the same mounting, should be able to achieve a resolution of the order of 0.1 milliarcseconds. Located

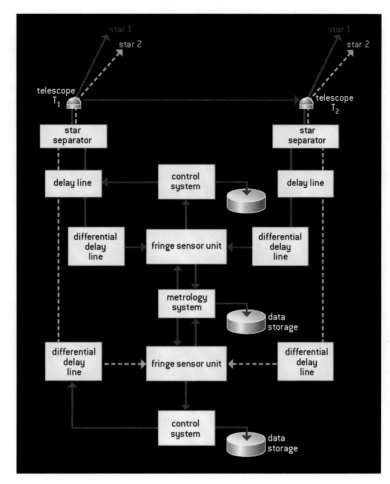

How the PRIMA interferometer works. To an accuracy of a few hundredths of a milliarcsecond, PRIMA will be able to detect giant exoplanets less than 5 AU from their stars, if those stars are within 15 parsecs from the Sun.

at the Very Large Telescope site in Chile, the Phase Referenced Imaging and Microarcsecond Astrometry (PRIMA) facility on the VLTI interferometer aims at a resolution of just a few hundredths of a milliarcsecond. Such capabilities should enable the detection of giant exoplanets of the size of Uranus, if they are located within 5 AU of their stars (assuming that those stars are within 15 parsecs of the Sun).

Further into the future more accurate measurements will be beamed to us from space. Some time after 2010, two astrometrically oriented space missions will be launched: Gaia (European Space Agency) and SIM (NASA). Both these missions will aim for angular resolutions of the order of a few microarcseconds. If they fulfil their promise they will be capable of detecting not only giant exoplanets, but also the exoEarths for which we have searched so long.

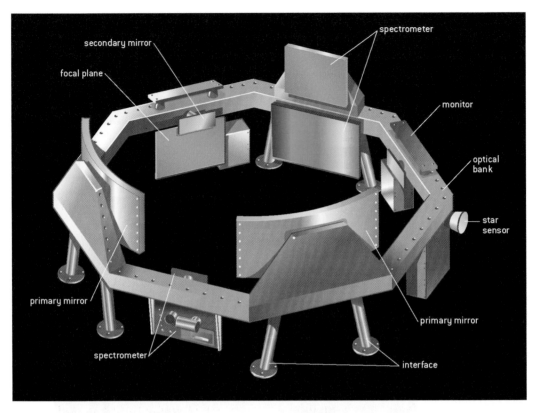

The optical bench of the Gaia space mission. (Courtesy European Space Agency.)

7.4 EXOEARTHS IN TRANSIT

COROT and Kepler

As we have already remarked, the transit of a giant exoplanet involves a dimming of starlight of the order of about 1%. This may be observable from the Earth, as has been shown by the detection of the exoplanet HD 209458 b. However, it would be vain to hope to detect the transit of an Earth-sized exoplanet with ground-based instruments, as the accuracy required in the measurement of the stellar flux (to a level better than 10^{-4}) is not possible over periods of hours. For this reason, astronomers on both sides of the Atlantic are turning their thoughts towards space missions.

The French COROT mission, conceived and developed by the Centre National des Etudes Spatiales (CNES) with international partners, involves a 25-cm telescope. This carries a wide-field camera with four ultrasensitive and highly stable detectors. This satellite – launched in December 2006 – is in a polar orbit

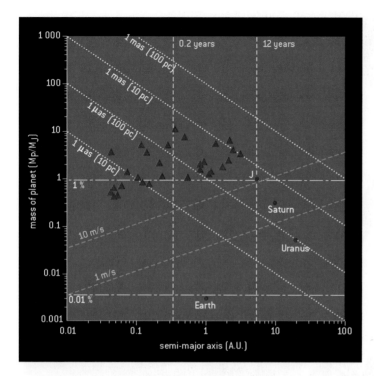

Detectability limits of planetary search methods. The red dots represent planets of the solar system, and the red triangles a sample of the extrasolar planets discovered to date. The white dotted lines show boundaries of detectability for astrometric methods (the domains are below the lines), and the orange dashed lines show boundaries of detectability for radial velocity methods. The yellow lines indicate photometric occultation.

around the Earth, at an altitude of approximately 900 km. This mission has two aims: the study of the internal structure of stars through the observation of their modes of oscillation; and the search for exoplanetary transits through systematic observation of about 12,000 stars. The COROT programme should result in the detection of not only many giant exoplanets, but also dozens of exoEarths, with diameters 1–2 times that of the Earth. COROT will also be able to detect possible rings or satellites of giant exoplanets during transits. The mission is due to last for about 2½ years. The Kepler project, now being developed by NASA, is even more ambitious. It is entirely dedicated to the search for exoplanets, by observing transits. The principle is identical to that of COROT, but the greater diameter of Kepler's telescope (1 metre) and its wide field of view (10° × 10°) may assist in the detection of possibly hundreds of exoEarths. Kepler is due to be launched by about 2008.

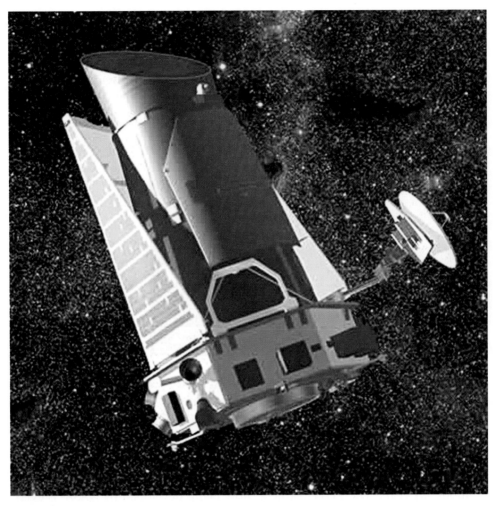

An artist's impression of NASA's Kepler mission. Like the COROT mission, Kepler is designed to seek out exoEarths by the transit method.

7.5 SEEING EXOPLANETS AT LONG LAST

What is the future for direct imaging?

Why is it so difficult to obtain a direct image of an exoplanet? The problem lies with the low contrast between the light reflected from the exoplanet and that of its star. As seen from the solar system, the exoplanet is so close to the star that its 'light' is lost in the brilliance of its neighbour. From a distance of 10 parsecs, the visible-light flux of the Earth (at 0.1 arcseconds from the Sun) represents only a

The first direct (false-colour) images of a brown dwarf (Gliese 229 B), obtained by the Hubble Space Telescope.

A four-quadrant coronagraph mounted on one of the four VLT telescopes in Chile. Coronagraphy enables the masking of the central star, enhancing the image of a nearby planet.

few billionths of the light of the Sun. In the thermal infrared the situation is less unfavourable: the planet emits its own thermal flux, which depends on its temperature. In the case of the Earth the contrast is now in the region of 10^{-6} at 10 μm. Let us apply this calculation to exoplanets: a Pegasid, compared with its star (and at 0.005 arcseconds from it) offers a contrast of 10^{-6} in visible light, while the value for a giant exoplanet at 5 AU (or 0.5 arcseconds from its star) is 10^{-7} (comparable to an exoEarth at 1 AU).

Therefore, if the exoplanet is near its star, the contrast of its flux compared to that of the star is higher, increasing the chances of observation; but this flux will still be drowned by starlight, obstructing the view. So, the observer is faced with the problem of finding a compromise between these two parameters. Observation of these unfavourable contrasts requires powerful measuring tools, and 'seeing' planets close to their stars requires a high degree of angular resolution. Such a venture seems possible with today's big telescopes, if the central stars are not too bright. In 2005, for example, an ESO team, using the deep survey 2Mass infrared equipment, secured and confirmed the first image of an exoplanet, 55 AU from a small, cool star 70 parsecs away.

Current projects can be divided into two main groups, according to the wavelengths at which they work. In the visible and the near-infrared (up to approximately 2 μm), a telescope of as much as 5 metres in diameter (probably the planned James Webb Space Telescope) will theoretically be able to observe, at the limit of diffraction, an exoplanet at 0.1 AU from a star 10 parsecs away. In order to distinguish a contrast of the order of 10^{-9}, sophisticated coronagraphical techniques will have to be employed. Research and development of such techniques are currently under way. Methods like these will probably have to be

Will exoEarths be detected from the ground? It is not totally impossible to detect and study exoEarths with ground-based spectroscopic techniques, but it will indeed be a very difficult undertaking. As exoplanets have such a weak signal, a very large collecting surface is needed. If a single telescope is used it has to be very big. Because exoplanets are very close to their stars, all atmospheric turbulence has to be corrected for, using adaptive optics, and all the resolving power of the telescope. Resolution, of course, depends on the size of the telescope: the bigger the better for this purpose. Calculations indicate that the detection of exoEarths by spectroscopy requires an aperture of more than 60 metres, which calls for quite a leap in technology (and cost!), as the biggest telescopes currently in use have apertures of 10 metres. More difficult even than the construction of such a telescope will be the creation of the adaptive optics, indispensable to the study of exoEarths. Several telescope projects, involving instruments with apertures of 30–100 metres, are currently being discussed in both Europe and the United States. The illustration is an artist's impression of OWL – the Overwhelmingly Large Telescope.

restricted to the detection of giant exoplanets if carried out by ground-based instruments; indeed, this will be the particular task of the ESO's VLT-PF/SPHERE (Planet Finder) project. This option is also being considered as part of the American Terrestrial Planet Finder (TPF) mission.

In the thermal infrared (5–30 μm) it is not possible, with a single telescope, to attain the necessary limit of diffraction (0.1 arcseconds) to separate an exoEarth from its star at a distance of 10 parsecs. At 10 μm an aperture of 25 metres would be needed. Until the new generation of Extremely Large Telescopes (ELT) are enabled, astronomers will rely on interferometry.

7.6 DETECTING EXOEARTHS WITH INTERFEROMETRY

Darwin and Terrestrial Planet Finder: awaiting direct images

Let us now return to the domain of the thermal infrared. Here, the contrast between the brightnesses of planet and star is much more favourable than in the visible. Again, with only one telescope there is no hope of securing a direct image of an exoplanet, unless that instrument has a diameter of several tens of metres – and such telescopes are still some decades away.

Once more, interferometry is required. The principle of interferometry, very simply put, is to replace one large telescope with two smaller ones at separate locations. They both observe the source to be studied, and the radiation from the sources is recombined to create interference fringes. Processing this information allows observers to realise an image of the source, along an axis on the sky in the same orientation as the line between the two telescopes. It can be shown that it is possible to achieve an angular resolution equivalent to that of a telescope of diameter equal to the distance between the two telescopes. To obtain a complete image of the source, at least three (unaligned) telescopes are needed – the aperture synthesis method. The technique of interferometry is currently used in radio astronomy.

How is this powerful technique applied in the search for exoEarths? When observing a companion body which is not only very near its star but also much fainter, the stellar signal has to be measured with very sophisticated equipment. To this end, an ingenious concept has been developed: Bracewell interferometry – the principle of which is as follows. In classical interferometry the signals from the two telescopes are combined to create a central interference fringe of

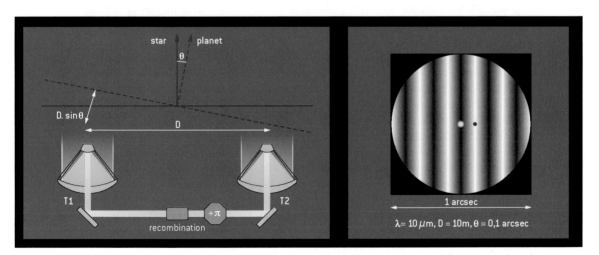

Principle of the Bracewell interferometer.

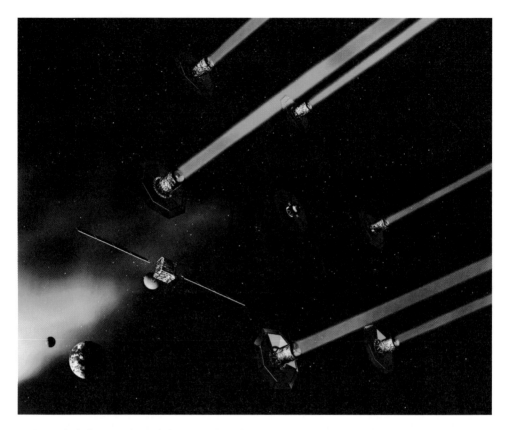

An artist's impression of the Darwin mission – a flotilla of six telescopes with apertures of more than 1 metre, working together as an interferometer, and positioned at about 5 AU from the Sun. This will enable the reconstitution of images of planets orbiting other stars. A more recent concept includes three telescopes of 3 metres and a communications hub.

maximum intensity along the line of sight (in the direction of the star observed). The neighbouring region, wherein the target exoplanet lies, corresponds to a negative fringe, where the received signal is zero. It is possible to introduce a 'phase shift of π' into the beam arriving at one of the telescopes, in such a way that the interferences become destructive in the direction of the star and positive in its vicinity. The flux of the star is thereby blocked, in the manner of a coronagraph. This is known as nulling interferometry.

The ambitious Darwin mission, now on the drawing board at ESA, is based around this method. Darwin will consist of a flotilla of three space telescopes, each 3 metres in diameter, together with a fourth spacecraft to serve as a communications hub. The ensemble will have to be located at the L2 Lagrangian point on the Earth–Sun axis above the night side of the Earth. This is a particularly convenient point for multiple satellites, because it is in a position of equilibrium in the Earth–Sun gravity field. Meanwhile, NASA is considering a similar concept as an option for its TPF mission (see p. 12). Given the associated technological challenges and high costs, it is probable that the two projects will be fused into one, which could be ready by the 2020 horizon. What might Darwin/TPF bring us? First, images of planetary systems which will probably already have been detected indirectly. Second, infrared spectra of these exoEarths – spectra which will perhaps carry the signature of ozone.

Appendix 1

The eight planets of the solar system

Name	Semimajor axis (AU)	Period of revolution (years)	Eccentricity (e)	Equatorial diameter (D_{Earth})	Mass (M_{Earth})
Terrestrial planets					
Mercury	0.387	0.241	0.206	0.382	0.055
Venus	0.723	0.615	0.007	0.949	0.815
Earth	1.000	1.000	0.017	1.000	1.000
Mars	1.524	1.881	0.093	0.532	0.107
Giant planets					
Jupiter	5.203	11.856	0.048	11.21	317.9
Saturn	9.537	29.424	0.054	9.45	95.16
Uranus	19.191	83.747	0.047	4.00	14.53
Neptune	30.069	163.273	0.009	3.88	17.14

Appendix 2

The first 200 extrasolar planets

Name	Mass (M_J)	Period (years)	Semimajor axis (AU)	Eccentricity (*e*)
14 Her b	4.74	1,796.4	2.8	0.338
16 Cyg B b	1.69	798.938	1.67	0.67
2M1207 b	5		46	
47 Uma b	2.54	1,089	2.09	0.061
47 Uma c	0.79	2,594	3.79	0
51 Peg b	0.468	4.23077	0.052	0
55 Cnc b	0.784	14.67	0.115	0.0197
55 Cnc c	0.217	43.93	0.24	0.44
55 Cnc d	3.92	4,517.4	5.257	0.327
55 Cnc e	0.045	2.81	0.038	0.174
70 Vir b	7.44	116.689	0.48	0.4
AB Pic b	13.5		275	
BD-10 3166 b	0.48	3.488	0.046	0.07
ε Eridani b	0.86	2,502.1	3.3	0.608
γ Cephei b	1.59	902.26	2.03	0.2
GJ 3021 b	3.32	133.82	0.49	0.505
GJ 436 b	0.067	2.644963	0.0278	0.207
Gl 581 b	0.056	5.366	0.041	0
Gl 86 b	4.01	15.766	0.11	0.046
Gliese 876 b	1.935	60.94	0.20783	0.0249
Gliese 876 c	0.56	30.1	0.13	0.27
Gliese 876 d	0.023	1.93776	0.0208067	0
GQ Lup b	21.5		103	
HD 101930 b	0.3	70.46	0.302	0.11
HD 102117 b	0.14	20.67	0.149	0.06
HD 102195 b	0.488	4.11434	0.049	0.06
HD 104985 b	6.3	198.2	0.78	0.03

Name	Mass (M$_J$)	Period (years)	Semimajor axis (AU)	Eccentricity (*e*)
HD 106252 b	6.81	1,500	2.61	0.54
HD 10647 b	0.91	1,040	2.1	0.18
HD 10697 b	6.12	1,077.906	2.13	0.11
HD 107148 b	0.21	48.056	0.269	0.05
HD 108147 b	0.4	10.901	0.104	0.498
HD 108874 b	1.36	395.4	1.051	0.07
HD 108874 c	1.018	1,605.8	2.68	0.25
HD 109749 b	0.28	5.24	0.0635	0.01
HD 111232 b	6.8	1,143	1.97	0.2
HD 114386 b	0.99	872	1.62	0.28
HD 114729 b	0.82	1,131.478	2.08	0.31
HD 114762 b	11.02	83.89	0.3	0.34
HD 114783 b	0.99	501	1.2	0.1
HD 117207 b	2.06	2,627.08	3.78	0.16
HD 117618 b	0.19	52.2	0.28	0.39
HD 118203 b	2.13	6.1335	0.07	0.309
HD 11964 b	0.11	37.82	0.229	0.15
HD 11977 b	6.54	711	1.93	0.4
HD 121504 b	0.89	64.6	0.32	0.13
HD 122430 b	3.71	344.95	1.02	0.68
HD 12661 b	2.3	263.6	0.83	0.35
HD 12661 c	1.57	1,444.5	2.56	0.2
HD 128311 b	2.18	448.6	1.099	0.25
HD 128311 c	3.21	919	1.76	0.17
HD 130322 b	1.08	10.724	0.088	0.048
HD 13189 b	14	471.6	1.85	0.28
HD 134987 b	1.58	260	0.78	0.24
HD 136118 b	11.9	1,209	2.3	0.37
HD 141937 b	9.7	653.22	1.52	0.41
HD 142 b	1	337.112	0.98	0.38
HD 142022 A b	4.4	1,923	2.8	0.57
HD 142415 b	1.62	386.3	1.05	0.5
HD 147513 b	1	540.4	1.26	0.52
HD 149026 b	0.36	2.8766	0.042	0
HD 149143 b	1.33	4.072	0.053	0.016
HD 150706 b	1	264	0.82	0.38
HD 154857 b	1.8	398.5	1.11	0.51
HD 160691 b	1.67	654.5	1.5	0.31
HD 160691 c	3.1	2,986	4.17	0.57
HD 160691 d	0.044	9.55	0.09	0
HD 16141 b	0.23	75.56	0.35	0.21
HD 162020 b	13.75	8.428198	0.072	0.277
HD 164922 b	0.36	1,155	2.11	0.05

Name	Mass (M$_J$)	Period (years)	Semimajor axis (AU)	Eccentricity (*e*)
HD 168443 b	7.2	58.116	0.29	0.529
HD 168443 c	17.1	1,739.5	2.87	0.228
HD 168746 b	0.23	6.403	0.065	0.081
HD 169830 b	2.88	225.62	0.81	0.31
HD 169830 c	4.04	2,102	3.6	0.33
HD 177830 b	1.28	391	1	0.43
HD 178911 B b	6.292	71.487	0.32	0.1243
HD 179949 b	0.98	3.092	0.04	0.05
HD 183263 b	3.69	634.23	1.52	0.38
HD 187085 b	0.75	986	2.05	0.47
HD 187123 b	0.52	3.097	0.042	0.03
HD 188015 b	1.26	456.46	1.19	0.15
HD 188753A b	1.14	3.3481	0.0446	0
HD 189733 b	1.15	2.218573	0.0313	0
HD 190228 b	4.99	1,127	2.31	0.43
HD 190360 b	1.502	2,891	3.92	0.36
HD 190360 c	0.057	17.1	0.128	0.01
HD 192263 b	0.72	24.348	0.15	0
HD 195019 b	3.43	18.3	0.14	0.05
HD 196050 b	3	1,289	2.5	0.28
HD 196885 b	1.84	386	1.12	0.3
HD 19994 b	2	454	1.3	0.2
HD 202206 b	17.4	255.87	0.83	0.435
HD 202206 c	2.44	1,383.4	2.55	0.267
HD 20367 b	1.07	500	1.25	0.23
HD 2039 b	4.85	1,192.582	2.19	0.68
HD 20782 b	1.8	585.86	1.36	0.92
HD 208487 b	0.45	123	0.49	0.32
HD 209458 b	0.69	3.52474859	0.045	0.07
HD 210277 b	1.24	435.6	1.097	0.45
HD 212301 b	0.45	2.457	0.036	0
HD 213240 b	4.5	951	2.03	0.45
HD 216435 b	1.49	1,442.919	2.7	0.34
HD 216437 b	2.1	1,294	2.7	0.34
HD 216770 b	0.65	118.45	0.46	0.37
HD 217107 b	1.37	7.1269	0.074	0.13
HD 217107 c	2.1	3,150	4.3	0.55
HD 219449 b	2.9	182	0.3	
HD 222582 b	5.11	572	1.35	0.76
HD 224693 b	0.71	26.73	0.233	0.05
HD 23079 b	2.61	738.459	1.65	0.1
HD 23596 b	7.19	1,558	2.72	0.314
HD 2638 b	0.48	3.4442	0.044	0

Name	Mass (M$_J$)	Period (years)	Semimajor axis (AU)	Eccentricity (e)
HD 27442 b	1.28	423.841	1.18	0.07
HD 27894 b	0.62	17.991	0.122	0.049
HD 28185 b	5.7	383	1.03	0.07
HD 30177 b	9.17	2,819.654	3.86	0.3
HD 330075 b	0.76	3.369	0.043	0
HD 33283 b	0.33	18.179	0.168	0.48
HD 33564 b	9.1	388	1.1	0.34
HD 33636 b	9.28	2,447.292	3.56	0.53
HD 34445 b	0.58	126	0.51	0.4
HD 3651 b	0.2	62.23	0.284	0.63
HD 37124 b	0.61	154.46	0.53	0.055
HD 37124 c	0.683	2,295	3.19	0.2
HD 37124 d	0.6	843.6	1.64	0.14
HD 37605 b	2.3	55	0.25	0.677
HD 38529 b	0.78	14.309	0.129	0.29
HD 38529 c	12.7	2,174.3	3.68	0.36
HD 39091 b	10.35	2,063.818	3.29	0.62
HD 40979 b	3.32	267.2	0.811	0.23
HD 41004 A b	2.3	655	1.31	0.39
HD 4203 b	1.65	400.944	1.09	0.46
HD 4208 b	0.8	812.197	1.67	0.05
HD 4308 b	0.047	15.56	0.114	0
HD 45350 b	1.79	890.76	1.92	0.778
HD 46375 b	0.249	3.024	0.041	0.04
HD 47536 b	4.96	712.13	1.61	0.2
HD 49674 b	0.11	4.948	0.0568	0.16
HD 50499 b	1.71	2,582.7	3.86	0.23
HD 50554 b	4.9	1,279	2.38	0.42
HD 52265 b	1.13	118.96	0.49	0.29
HD 59686 b	5.25	303	0.911	0
HD 62509 b	2.9	589.64	1.69	0.02
HD 63454 b	0.38	2.81782	0.036	0
HD 6434 b	0.48	22.09	0.15	0.3
HD 65216 b	1.21	613.1	1.37	0.41
HD 66428 b	2.82	1,973	3.18	0.465
HD 68988 b	1.9	6.276	0.071	0.14
HD 69830 b	0.033	8.667	0.0785	0.1
HD 69830 c	0.038	31.56	0.186	0.13
HD 69830 d	0.058	197	0.63	0.07
HD 70642 b	2	2,231	3.3	0.1
HD 72659 b	2.96	3,177.4	4.16	0.2
HD 73256 b	1.87	2.54858	0.037	0.03
HD 73526 b	2.9	188.3	0.66	0.19

Name	Mass (M_J)	Period (years)	Semimajor axis (AU)	Eccentricity (e)
HD 73526 c	2.5	377.8	1.05	0.14
HD 74156 b	1.86	51.643	0.294	0.636
HD 74156 c	6.17	2,025	3.4	0.583
HD 75289 b	0.42	3.51	0.046	0.054
HD 76700 b	0.197	3.971	0.049	0.13
HD 80606 b	3.41	111.78	0.439	0.927
HD 81040 b	6.86	1,001.7	1.94	0.526
HD 82943 b	1.75	441.2	1.19	0.219
HD 82943 c	2.01	219	0.746	0.359
HD 83443 b	0.41	2.9853	0.04	0.08
HD 8574 b	2.23	228.8	0.76	0.4
HD 86081 b	1.5	2.1375	0.039	0.008
HD 88133 b	0.22	3.41	0.047	0.11
HD 89307 b	2.73	3,090	4.15	0.27
HD 89744 b	7.99	256.605	0.89	0.67
HD 92788 b	3.86	377.7	0.97	0.27
HD 93083 b	0.37	143.58	0.477	0.14
HD 99109 b	0.502	439.3	1.105	0.09
HD 99492 b	0.109	17.0431	0.1232	0.254
HIP 14810 b	3.84	6.674	0.0692	0.148
HIP 14810 c	0.951	113.8267	0.458	0.2806
HIP 75458 b	8.82	511.098	1.275	0.7124
HR 810 b	1.94	311.288	0.91	0.24
OGLE-05-071L b	0.9	2,900	1.8	
OGLE-05-169L b	0.04	3,300	2.8	
OGLE-05-390L b	0.017	3,500	2.1	
OGLE-TR-10 b	0.54	3.101269	0.04162	0
OGLE-TR-111 b	0.53	4.0161	0.047	0
OGLE-TR-113 b	1.32	1.4324757	0.0229	0
OGLE-TR-132 b	1.19	1.689857	0.0306	0
OGLE-TR-56 b	1.45	1.2119189	0.0225	0
OGLE235-MOA53 b	2.6		5.1	
PSR 1257+12 b	0	25.262	0.19	0
PSR 1257+12 c	0.014	66.5419	0.36	0.0186
PSR 1257+12 d	0.012	98.2114	0.46	0.0252
PSR 1620-26 b	2.5	36,525	23	
ρ CrB b	1.04	39.845	0.22	0.04
SCR 1845 b	8.5		4.5	
τ Boo b	3.9	3.3135	0.046	
TrES-1	0.61	3.030065	0.0393	0.135
υ And b	0.69	4.617	0.059	0.012
υ And c	1.89	241.5	0.829	0.28
υ And d	3.75	1,284	2.53	0.27
XO-1 b	0.9	3.941534	0.0488	

Glossary

Accretion The process by which matter falls together to form solid bodies such as planetesimals, cometary nuclei, asteroids, planets and so on.

Albedo The fraction of incident light reflected by a body.

Aphelion In an orbit, the furthest point from the Sun.

Arcminute A unit of angular measurement: $1° = 60$ arcmin (minutes of arc), and 1 arcmin = 60 arcsec (seconds of arc). The angular diameter of the Moon is about 30 arcmin.

Asteroid belt The main belt of known asteroids, between the orbits of Mars and Jupiter.

Asteroids (minor planets) Solar system objects orbiting the Sun, of sizes between a fraction of a kilometre and nearly 1,000 kilometres. Most of them are in the asteroid belt, between Mars and Jupiter.

Astrometry A method by which the proper motions of stars are measured. It has been used unsuccessfully to search for possible companions of nearby stars, but the required precision has yet to be achieved.

Astronomical Unit (AU) The mean distance between the Earth and the Sun – about 150 million km. This unit is used to measure distances within the solar system.

Atmosphere A gaseous envelope surrounding a celestial body. Atmospheres are found around planets and some satellites, and comets may develop them through outgassing as they approach the Sun or their central star.

Black body A body which absorbs all incident radiation (albedo = 0).

Brown dwarf A low-mass star (less than 0.08 of the mass of the Sun, or 74 Jupiter masses), too cool to trigger within itself the nuclear reactions which will transform hydrogen into helium. Nuclear reactions in brown dwarfs involve deuterium, and last for perhaps ten million years only. If the mass is less than 13 Jupiter masses, the temperature will not even be sufficient to trigger deuterium reactions, and it is an exoplanet rather than a star.

Circumstellar region The immediate environment of a star within which its radiation and stellar wind interact with the rest of its protoplanetary disk and the nearby interstellar medium.

Coma A cometary envelope containing gas, ices and dust ejected from the comet's nucleus as a result of solar radiation.

Comet Comets are objects formed from ice and dust, a few kilometres in diameter. When they approach the Sun, outgassing of volatile elements causes a coma and a tail to develop, sometimes offering a spectacular sight in the sky. The difference between comets and asteroids is not very well defined: certain asteroids may be comets which have lost their reserves of ice after many returns to the vicinity of the Sun.

Cometary nucleus The solid part of a comet, its size ranging from a few kilometres to tens of kilometres, composed of ices and dust grains.

Cometary tail A collection of large cometary dust particles diffusing along the orbit.

Cosmic rays Very energetic particles emanating from the Sun and other stars in our Galaxy.

Crater A circular depression on the surface of a celestial body. Craters on the planets and satellites of the solar system may result from volcanic activity or infalling meteorites.

Debris disk A disk of dust surrounding main-sequence stars. The presence of such disks, essentially composed of dust, may indicate the presence of extrasolar planets orbiting the star.

Differentiation In the solar system, differentiated bodies are those which have undergone internal transformations since they formed. Within the Earth, for example, heating has caused heavy elements to collect at the centre to form a metallic core, leaving a lighter crust above. Low-mass objects such as comets, consisting of conglomerated dust and ice, are not differentiated. Asteroids may show some differentiation, due to the energy liberated by radioactive elements within them.

Doppler Effect The shift in frequency observed in the spectrum of a star moving with respect to the observer. The shift is towards lower frequencies (the red end of the spectrum) if the source is receding, and towards higher frequencies (the blue end) if it is approaching. The Doppler effect forms the basis of the velocimetric method for the detection of exoplanets.

Drake equation An equation, ascribed to Frank Drake and Carl Sagan, that attempts to quantify the probability of detecting an extraterrestrial civilisation. Among the main factors in the equation are the number of exoplanets capable of harbouring life, and the fraction of those planets where a civilisation becomes technologically advanced.

Eccentricity A parameter characterising the shape of an orbit. The eccentricity of a circular orbit is 0; of an ellipse, 0–1; of a parabola, 1; and of an hyperbola, >1.

Ecliptic The plane in which the Earth travels around the Sun. The orbits of all the planets are very close to the ecliptic plane.

ExoEarth An Earth-like planet beyond the solar system. Although searches for these objects are being undertaken, current observational techniques have not yet been able to reveal any.

Extrasolar planet (exoplanet) A planet orbiting a star other than the Sun. More than 200 exoplanets are now known.

Galaxy, The A collection of stars, of which one is the Sun. Its stars are distributed in spiral arms within a disk which we see edge-on (the Milky Way) from our viewpoint near the Sun. The Sun is 8 kiloparsecs from the Galactic centre, which lies in the direction of the constellation of Sagittarius.

Giant planets Jupiter, Saturn, Uranus and Neptune. They are very massive and not very dense, and their surfaces are hidden by atmospheres composed of hydrogen, helium and, in lesser quantities, methane and ammonia. They have ring systems and many satellites.

Gravitational microlens If a distant star is occulted by another object outside the solar system (such as an extrasolar planet or a brown dwarf), the radiation from that star is modified. This is not caused by atmospheric refraction (as is the case when the occultation is by a planet in the solar system), but by the curvature of space induced by the occulting object, as predicted by the theory of relativity. This phenomenon is known as gravitational microlensing by analogy with the larger-scale phenomenon of gravitational lensing, when the occulting object and the more distant object are both galaxies.

Greenhouse effect The mechanism by which a planet's surface is warmed through the absorption of infrared radiation emitted by that surface into the atmosphere above it. The spectroscopic properties of carbon dioxide, water vapour and methane cause them to be particularly effective greenhouse gases. In the absence of a regulatory mechanism, the greenhouse effect tends to accelerate, as has happened in the case of Venus.

Habitability zone The region in the environment of a given star where any planet present might be able to maintain a liquid ocean. For a Sun-like star, this zone is situated at around 1 AU. For a less massive (and therefore less bright) star, it would be nearer, and for a more massive (brighter) star, it would be further away.

Heavy element Any chemical element with an atomic mass equal to or greater than that of carbon.

Heliocentric distance Distance from the Sun, usually expressed in Astronomical Units (AU).

Heliocentric (Copernican) system A representation of the Universe according to which the planets revolve around the Sun. The heliocentric system became widely known due to Nicolaus Copernicus's *De Revolutionibus Orbium Coelestium*, published in 1543.

Hot Jupiters Extrasolar planets of similar mass to Jupiter, which orbit close to their stars. Their temperatures are therefore much higher than that of Jupiter.

Ice line Defined by the condensation temperature of water and therefore at a certain distance from the Sun, the ice line marks the boundary between the gaseous component of protosolar molecules associated with carbon, nitrogen and oxygen (to the Sunward side of the boundary) and their solid component (beyond the boundary). It has played a major part, among other factors, in the separation of the planets into two distinct classes: terrestrial planets and gas giants.

Imaging Observing a celestial body at a given wavelength, using a two-dimensional receiver. The first receivers were photographic plates working in visible light. Nowadays cameras are used at wavelengths from the visible to the submillimetre regions.

Infrared (IR) Spectral domain of wavelengths from 100 μm to 800 nm. With the exception of a few windows accessible from high-altitude observatories, infrared radiation is studied using aircraft, balloons and satellites.

Interferometry A method of detecting celestial bodies by combining signals from a single source, using more than one telescope, enabling the image of the source to be established with considerable accuracy. Originally a radio technique, it is nowadays used also in the infrared and visible domains.

Interstellar cloud A cloud of gas and dust which, together with stars, is a major constituent of galaxies. Some interstellar clouds are diffuse and of low density, and are essentially composed of atoms. Some clouds are much denser, such as the dark nebulae which block out the light of the stars behind them. Dense, dark clouds are rich in molecules, and are therefore called molecular clouds.

Interstellar grains Small solid particles present in clouds in interstellar space, sometimes incorporated intact into comets.

Isotope An atom of a given element having the normal number of protons and electrons (and therefore the same chemical properties), but a different number of neutrons. The isotopic ratios of an element provide an indication of the history of its formation in the Universe. Some isotopes are unstable and disintegrate over a period which may be very long (sometimes billions of years) or very short, and they can therefore be used as 'clocks', The isotopic composition of meteorites has provided information on the age of the solar system, formed 4.55 billion years ago (give or take 100 million years).

Isotopic dating A method by which the age of solar system bodies may be determined, based on the disintegration of certain long-lived radioactive elements present in the objects.

Jet An ejection of matter from a celestial body (such as a star or cometary nucleus) in a specific direction.

Kepler's laws Johann Kepler's three laws – the first two of which were published in 1609 and the third in 1619 – are the empirical laws describing the movements of the planets around the Sun. They form the basis of the law of universal gravitation proposed by Isaac Newton in 1687, and apply to any celestial body in an orbit around a more massive body.

Kelvin A measurement of temperature. Absolute zero is 0 K (−273°.15 C).

Kuiper Belt (Edgeworth–Kuiper Belt) A collection of planetesimals (asteroids and cometary nuclei) orbiting beyond Neptune, at distances between 30 and 100 AU. The Kuiper Belt is thought to be the source of short-period comets of the Jupiter family.

Light-year The distance that light travels in a vacuum in one year (63,000 AU).

Metallicity The percentage of heavy elements in the chemical composition of a star. In the case of cosmic abundances, the relative mass of heavy elements (elements heavier than helium) is 2% of the total. Certain stars exhibit an excess of heavy elements relative to this mean value, and the metallicity of the star is an indicator of the excess.

Meteor A luminous phenomenon caused by the passage of a meteoroid through the atmosphere; commonly referred to as a 'shooting star'. A group of meteors is known as a shower.

Meteorite An extraterrestrial object which has reached the Earth's surface. Most meteorites originate in asteroids, while some come from the Moon and Mars. The smallest meteorites (micrometeorites) are probably of cometary origin.

Migration The possible movement of a planet within a protoplanetary disk. Instabilities in the density of the disk cause the planet to move inwards towards the central star.

Minute of arc (see *arcminute*).

Multiple-planet system A system in which two or more extrasolar planets orbit a star.

Nucleosynthesis The train of nuclear reactions leading to the production of the different chemical elements. At the time of the primordial nucleosynthesis, completed a few minutes after the Big Bang, the lightest elements – hydrogen, deuterium, helium and lithium – were created. The heavier elements are synthesised by stars. In stellar nucleosynthesis, hydrogen is transmuted into helium, and so on to form carbon, nitrogen, oxygen, and other heavier elements.

Occultation A phenomenon involving a celestial object (a planet or satellite) passing directly in front of another. When a planet occults a star, the way in which the star's light is absorbed by the planet's atmosphere (if present) tells us about the composition of the atmosphere and the temperature profile according to altitude. If the planet has a ring system that passes in front of the star, the light of the star will appear to dim intermittently. The rings of Uranus and Neptune were discovered in this way.

Ocean planet A theoretical extrasolar planet, of a size between a terrestrial planet and a giant planet. Such extrasolar planets, comparable perhaps to Uranus (17 times the mass of the Earth), might have cores composed essentially of ice, which, if they migrated towards their central stars, might form oceans of liquid water. At present, ocean planets exist only in theory, but the discovery of extrasolar planets with masses between 14 and 20 Earth masses lends added interest to the hypothesis.

Oort Cloud A hypothetical cloud of comets uniformly spread at a distance of 20 000 to 100 000 AU from the Sun – a source of new comets.

Organic compound Molecules composed of carbon, hydrogen and (possibly) oxygen and nitrogen.

Osiris The extrasolar planet HD 209458 b – a Pegasid-type planet detected by both velocimetry and the transit method. Osiris is therefore one of the very rare extrasolar planets of which physical and geometrical parameters are known. It orbits its star at a distance of 0.045 AU, has a diameter 1.347 times that of Jupiter, and a mass 0.69 times that of Jupiter.

Outgassing The ejection of gas from a celestial body. In the case of the giant planets, outgassing is associated with sublimation of ices from the core due to heating.

Parsec (pc) The distance from the Earth to a star at which the semi-major axis of Earth's orbit subtends an angle of 1 arcsecond (1 parsec = 3.261633 light-years).

Pegasid A giant extrasolar planet orbiting very close to its star, at a distance of the order of 0.05 AU. Such extrasolar planets are named after the first extrasolar planet to be detected – 51 Pegasi b, which belongs to this category.

Perihelion In an orbit, the nearest point to the Sun.

Photodissociation The dissociation of a molecule due to radiation. In the case of a planetary atmosphere, photodissociation is caused by solar radiation.

Photometry A technique for measuring the intensity of radiation.

Planetesimal When the primordial nebula collapsed into a disk, collisions between dust grains within the interstellar cloud created tiny aggregates about 1 μm across: planetesimals. The next stage saw planetesimals growing as a result of collisions, and then drawing in material from around them (accretion) to become protoplanets.

Prebiotic Preceding the emergence of life. Prebiotic molecules are organic molecules essential to the chains of reactions leading to the formation of amino acids, constituents of proteins. Some of these molecules are present in the atmosphere of Titan, causing some astronomers to see this satellite of Saturn as a laboratory of prebiotic chemistry.

Primordial nebula/protosolar nebula A cloud of gas and dust which, having collapsed into a disc, gave birth to the Sun and its planets.

Protoplanet An embryonic planet in the process of formation.

Protoplanetary disk A disk of material resulting from the collapse of a rotating interstellar cloud. The matter collapses into a disk, and its central region falls together, eventually to form a star. Further out from the centre, planets gradually form by accretion and collisions between planetesimals.

Protostar An embryonic star formed from the contraction of material at the centre of a protoplanetary disk.

Pulsar A star emitting a characteristic radio signal of very brief pulses, of extreme regularity. The period of these signals may vary from milliseconds to seconds. These emissions come from a point on the star, which is rotating very rapidly and sending out radio waves in much the same way as the beams of a lighthouse. Pulsars are probably neutron stars, which may have the same mass as the Sun but diameters of only tens of kilometres. Anomalies detected in the periodicity of two pulsars suggest that there may be planets orbiting these objects.

Radial velocity That component of the velocity of a celestial body along a line of sight from the observer. This is the parameter measured when using the velocimetric method to detect exoplanets.

Radio The spectral domain of wavelengths greater than 100 μm. It embraces in particular the millimetre (1–10 mm) and submillimetre (0.1–1 mm) ranges, parts of which are inaccessible to Earth-based radio telescopes.

Red giant A very large star in a late stage of its evolution, having transformed its hydrogen and helium into heavy elements.

Refractory (molecules) Molecules able to remain solid at quite high temperatures; for example, silicates and metals.

Resonance A situation involving a body orbiting a more massive object, and being subjected to a periodic gravitational perturbation by another orbiting body. The periods of revolution of the two bodies will settle into a simple whole-number ratio (1:2, 3:2, 4:3...).

Revolution period The time taken for a celestial body to describe one complete orbit.

Ring A collection of solid particles orbiting a planet, and spread along its orbit in a uniform manner. The four giant planets of the solar system all have ring systems.

Rotation period The time taken for a celestial body to rotate once on its axis.

Satellite A celestial body orbiting a planet.

Single-planet system A system in which one extrasolar planet orbits a star.

Solar wind; stellar wind A continuous stream of energetic particles flowing from the Sun and other stars. The solar wind is essentially composed of electrons and protons travelling at speeds of about 400 km/s by the time they reach the Earth.

Spectrum The distribution of the radiation from a body as a function of wavelength. This distribution presents a maximum at a wavelength dependent upon the temperature of the source observed. In the case of the Sun, with a photospheric temperature of 5,770 K, this maximum is found in the visible-light domain at 0.5 μm. As the temperature of the source decreases, the maximum moves towards the infrared. In the case of the Earth (T = 300 K), it is found at around 13 μm.

Spectroscopy The analysis of the electromagnetic radiation emitted by a celestial body as a function of wavelength. It is used to identify the elements present in the source, by detection of emissions at certain wavelengths characteristic of the elements.

Star-field A collection of stars in a given area of sky as observed by instruments.

Supernova A late-stage massive star will have synthesised within it all the heavy elements up to and including iron. When the star contains only iron, it implodes as gravity dictates, and the outer layers are ejected into space: a supernova explosion.

T Tauri phase A particularly active period in the life of young stars, during which they exhibit considerable variations in brightness associated with the ejection of large quantities of matter. The T Tauri phase occurs during the first 10 million years of a star's life. The name derives from the star T Tauri, which is at present in this phase of activity.

Terrestrial planet Mercury, Venus, Earth and Mars. They are dense, relatively small, and have atmospheres (except Mercury) and surfaces accessible to observation. They have few or no satellites.

Tides Phenomena arising from the difference in the gravity field at two points in an object. When a small body is near a larger and more massive body, tidal effects may be strong enough to cause volcanic activity on the smaller body, or even break it up.

Transit (planetary) The passage of a planet in front of a star. Observing such phenomena is one way of detecting extrasolar planets. In its movement around a star, a planet may pass between the star and the Earth, causing a diminution of the star's brightness. If the star and the planet have radii respectively R and r, the relative diminution of the flux will equal $(r/R)^2$.

Trans-Neptunian object (TNO) An object orbiting the Sun at a distance of more than 30 AU, within the Kuiper Belt. Pluto – detected in 1930 – was the first TNO to be discovered. More than 1,000 of these objects have been discovered in the last fifteen years.

Ultraviolet (UV) The spectral domain of wavelengths below 350 nm. UV radiation shorter than 300 nm may be observed from the ground, although it is attenuated by the atmosphere. Beyond this wavelength, observation by rockets and satellites is necessary.

Velocimetry A method of indirectly detecting the presence of an extrasolar planet orbiting a star by measuring the effect of gravitational perturbations affecting the radial velocity of the star.

Visible The spectral domain of wavelengths between 350 and 800 nm, corresponding to what is seen by the human eye.

Volatile (molecule) A molecule which sublimates or condenses at a relatively low temperature; for example, molecules which constitute cometary ices.

White dwarf The type of star that represents the last stage in the evolution of a solar-type star. After the combustion of all its hydrogen and helium the star becomes a red giant, and then ejects its outer envelope, The result is a planetary nebula with, at its heart, a small and very dense star – a white dwarf.

Bibliography

Books

Bertout, C., *Mondes lointains, à la recherche d'autres systèmes solaires*, Éd. Flammarion, 2003.

Bordé, Pascal, *Y a-t-il d'autre planètes habitées dans l'Univers?*, Éd. Le Pommier, 2004.

Bruine, E., *Les Jupiters chauds*, Éd. Belfond, 2002. A romantic view of the discovery of the first extrasolar planets.

Deming, D. and Seager, S., 'Scientific frontiers in research on extrasolar planets', *Astronomical Society of the Pacific Conference Series*, **294**, 2003.

Dick, S.J., *La pluralité des mondes*, Éd. Actes Sud, 1989. A history of the question of extraterrestrial life.

Encrenaz, T., *Les planètes geantes*, Éd. Belin, 1996.

Encrenaz, T., *À la recherche de l'eau dans l'Univers*, Éd. Belin, 2004.

Encrenaz, T., *Système solaire, systèmes stellaires*, Éd. Dunod, 2005.

Fontenelle, B. de Bovier, *Entretiens sur la pluralité des mondes*, Éd. de l'Aube, 2005.

Forget, F., Costard, F. and Lognonné, P., *La planète Mars: Histoire d'un autre monde*, Éd. Belin, 2004.

Heidmann, J., *Intelligences extra-terrestres*, Éd. Odile Jacob, 1992.

Heidmann, J., Vidal-Madjar, A., Prantzos, N. and Reeves, H., *Sommes-nous seuls dans l'univers?*, Éd. Fayard, 2000.

Mayor, M. and Frei, P.-Y., *Les nouveaux mondes du cosmos*, Éd. du Seuil, 2001.

Nazé. Y., *Les couleurs de l'Univers*, Éd. Belin, 2005.

Prantzos, N., *Voyages dans le futur*, Éd. du Seuil, 1998.

Raulin-Cerceau, F., Léna, P. and Schneider, J., *Sur les traces du vivant: de la Terre aux étoiles*, Éd. Le Pommier, 2002.

Vidal-Madjar, A., *Il pleut des planètes*, Éd. Hachette, 1999.

Internet sites

Geneva Observatory extrasolar planet search programmes
    http://obswww.unige.ch/~udry/planet/planet.html
California and Carnegie Planet search
    http://exoplanets.org/
Jet Propulsion Laboratory
    http://planetquest.jpl.nasa.gov/
European Southern Observatory
    http://www.eso.org
Jean Schneider: L'incontournable encyclopédie des planetes extrasolaires
    http://vo.obspm.fr/exoplanetes/encyclo/f-encycl.html
Promenade dans le système solaire
    http://www.imcce.fr/page.php?nav=fr/ephemerides/astronomie/
    Promenade/debutweb.php
Les neuf planètes du système solaire
    http://www.neufplanetes.org/
De la planète rouge à l'origine de la vie, un site sur la planète Mars
    http://www.nirgal.net/
Vie extraterrestre (exobiology)
    http://www.exobio.cnrs.fr/
SETI Institute (Search for ExtraTerrestrial Intelligence)
    http://www.seti.org

Index

Printing: Mercedes-Druck, Berlin
Binding: Stein+Lehmann, Berlin